Famous Wisconsin Inventors & Entrepreneurs

Marv Balousek

Badger Books Inc.
Oregon, WI

ISBN 1-878569-87-2

Badger Books Inc.
P.O. Box 192
Oregon, WI 53575
Toll-free phone: (800) 928-2372
Fax: (608) 835-3638
Email: books@badgerbooks.com
Web site: www.badgerbooks.com

To Buddy and Kristin Ruhland

Contents

INTRODUCTION

The Wisconsin Idea certainly was some thing not in short supply since the mid-1800s. Perhaps it's no surprise that inventors and entrepreneurs from the Dairy State contributed greatly to the mechanical revolution that changed agriculture. But that mechanical bent also can be seen in inventions ranging from the automobile to the typewriter.

If there was a mecca for inventors in Wisconsin, it was Racine, where Jerome Increase Case built a farm implement empire and Chester Beach's small universal motors spawned a plethora of household appliances. Besides these famous Racine inventors, however, the city had many others such as William Didier, whose loss of a leg in an industrial accident inspired him to invent bendable artificial arms and legs.

In addition to Case, people who changed the face of agriculture included John Appleby, whose invention of the twine binder allowed more mechanized grain harvesting, and Stephen Babcock, whose simple butterfat test established a quality standard for milk.

It's no surprise that Wisconsin also produced renowned environmentalists like John Muir and

Increase Lapham. But these men also were inventors. Muir invented several devices as a young man, including one that sprang him out of bed in the morning, and Lapham is recognized as founder of the U.S. Weather Service.

William Harley and Arthur Davidson are among Wisconsin's most famous entrepreneurs for their company's domination of the motorcycle market for the past century. Ole Evinrude was inspired to invent a practical boat motor after an ice cream cone he bought for his girlfriend melted on his way back across a lake. But Wisconsin also produced many lesser known inventors such as the Rev. John Carhart, whose unique automobile called the Spark terrified people and horses with its noise, and Carl Eliason, who invented the snowmobile.

The state is represented in the fashion industry by Freeman Tripp, who ran a women's clothing store in Madison during the 1800s and invented corsets, hat pins and a spring-loaded wagon tongue that made the job easier for beasts of burden. Rolf Darbo was a more modern inventor who invented a sandwich cooler and the Insty Grill.

Seymour Cray of Eau Claire has been called the Thomas Edison of the computer industry and Brook Stevens of Milwaukee designed many popular items including the Oscar Mayer Wienermobile.

I also have profiled master entrepreneurs

Tommy Bartlett and Alex Jordan Jr., who built giant tourist attractions in Wisconsin Dells and near Spring Green.

Female inventors and entrepreneurs are in short supply in Wisconsin, but there are a few that are notable including Pleasant Rowland, who built the American Girl brand of dolls and books into a multi-million-dollar empire; Margarethe Meyer Schurz, who brought the kindergarten concept to the United States from Germany; and Catherine Taft Clark, who founded Brownberry Ovens.

I hope you enjoy reading about these inventors and entrepreneurs as much as I did discovering them.

— Marv Balousek

HARLEYS & DAVIDSONS

Two Milwaukee families established one of the world's most enduring businesses. The Harley-Davidson company survived many bumps in the road to become the nation's top motorcycle manufacturer. The company celebrated its 100th anniversary in 2003.

It was in 1903, the same year that Wilbur and Orville Wright flew the first airplane, that William Harley and Arthur Davidson built their first successful motorcycle. At the time, there wasn't anything unique about building a motorcycle. Dozens of companies were attaching motors to bicycles for motorized transportation. But the Harley-Davidson models would outlast them all.

William Harley was born in Milwaukee in 1880. His parents had come to America from England. Arthur Davidson was a childhood friend and both enjoyed tinkering with mechanical things. By the turn of the century, both young men worked for a Milwaukee manufacturing firm. William was a draftsman and Arthur was a pattern maker.

Their first motorcycle was not successful because the engine was too small and the bicycle frame wasn't strong enough. They got help from Ole Evinrude, who would make his mark in outboard boat motors, in designing the carburetor. They also were encouraged by their families. Arthur's older brothers, Walter, a railroad mechanic, and William, a railroad toolmaker, soon joined the effort. Davidson's father, who had been a carpenter in his native Scotland, built a 10-foot by 15-foot shed in the backyard that served as the first Harley-Davidson factory.

William Harley took engineering courses at the University of Wisconsin so he could gain the knowledge to build a better motorcycle. The first motorcycle was sold to a customer identified as Mr. Meyer, who drove it for six thousand miles before he sold it. The bike was still on the road a decade later and had gone more than 100,000 miles.

Just three motorcycles were built and sold in each of the first two years. By 1906, a new plant was built at what became the company's permanent location, and production had increased to fifty motorcycles. The Harley-Davidson machine was nicknamed the Silent Gray Fellow, a reference to its quiet muffler and nondescript color. The following year the Harley-Davidson factory produced more than 150 machines.

Walter Davidson, who became the now incorporated company's first president, was the best cyclist among the two families of brothers, and he soon developed an interest in racing. Spectators who watched the races were potential customers. The Harley-Davidson motorcycle was designed more as cheap transportation than a racing machine, but the bike was matched in races with Indian motorcycles, manufactured in Connecticut by George Hendee and Oscar Hedstrom. Although the Harley machine couldn't win speed races, Davidson won an endurance race from the Catskill Mountains to New York City and then around Long Island.

Wisconsin State Journal archives

William Harley

When seven people died at a New Jersey motorcycle race in 1912, it marked the end of speed racing, although cross country racing survived.

In 1911, Harley-Davidson began producing the 1000cc V-Twin and the twin-carburetor, exposed engine would become a Harley trademark. The company tried to promote its bikes as family vehicles with Dad on the bike and Mom and the kids riding in the sidecar. But Harley was no match for the marketing genius of Henry Ford. The first Model T Ford was priced at $850, but the price was cut in half within five years, making it about $100 more than a Harley. That $100 bought slightly more comfort and a vehicle that was useful in all kinds of weather. A half million Fords were on the road in 1910 and a decade later the number had increased to eight million. In 1920, Harley produced about 23,000 motorcycles.

Harley-Davidson also marketed its motorcycles as freight carriers with a freight box in front of the driver. Ads proclaimed: "Don't use tonnage trucks for poundage delivery." The company was more successful selling its bikes to police forces and dispatch couriers.

By this time, the field had narrowed and many motorcycle manufacturers were out of business. Remaining Harley-Davidson competitors included Indian motorcycles and the Excelsior made by Chicago bicycle manufacturer Ignaz Schwinn.

During World War I, Harley-Davidson's production was turned to the war effort, and the bikes were used in combat. The Army green color persisted a few years after the war. About 30,000 motorcycles were supplied to the U.S. and its allies. After the war, Harley-Davidson focused on marketing its sidecars in yet another attempt to make inroads on automobile sales. It also tried to make motorcycles more acceptable transportation for women, who had won the right to vote but continued to prefer automobiles. Despite those failures, a major part of the company's success over its motorcycle rivals was its emphasis on advertising and public relations.

Still a family business, Harley-Davidson survived the stock market crash of 1929 and the Great Depression, although only 3,300 motorcycles were produced in 1933. The company also tried to expand its overseas market, moving briefly into Japan before that country's relations soured with the U.S. leading up to World War II.

The second world war brought another opportunity for Harley to enhance its sales by supplying the military. An estimated 90,000 bikes were manufactured for military use during the war. Widespread military use of Harley motorcycles helped create a generation of riders when the soldiers came home. Some of those who had ridden Harleys in the military bought their own bikes as civilians.

Walter Davidson, who had steered the company's growth during its formative years, died in 1942 and William Harley, who helped create those early motorcycles, died the following year. Walter and Arthur Davidson's brother William had died in 1937. By this time, a younger generation of Harleys and Davidsons had taken their places in the company's management. Walter Davidson's son, William H., succeeded his father as company president in 1942. Arthur

Davidson, the last of the original four company founders, died in an automobile crash in 1950 at age 69.

The postwar period spawned outlaw motorcycle gangs, which both plagued the Harley-Davidson's image and provided a customer niche. The period also brought more competition in the form of imported British motorcycles like Triumphs and Nortons. The lighter bikes from overseas stimulated Harley to overhaul its product line, and also meant the death of the Indian motorcycle manufacturer in 1953, leaving Harley the only U.S. motorcycle manufacturer. The same year, Harley-Davidson marked its 50th anniversary with the introduction of new models. The company's accessories catalog — caps, jackets, etc. — had evolved into an important aspect of the business.

Harley-Davidson executives began a long political war against the lower tariffs that the U.S. charged foreign manufacturers. While Harley paid a 40 percent tariff to export its motorcycles to other countries, foreign manufacturers paid 8 percent to bring their products to the U.S.

By the 1960s, books and movies had popularized roving motorcycle gangs and especially Hell's Angels. While their close association with Harley bikes tarnished the company's image in suburban America, it enhanced the image among coming-of-age baby boomers who rebelled against their parents' values. The arrival of Japanese imports like Honda, Kawasaki, and Yamaha nearly toppled Harley-Davidson, which continued to battle the tariff inequities. Suburban riders seemed to prefer the smaller Japanese bikes while Harley sold its motorcycles to gang members and police forces.

By this time, Harley-Davidson had drifted far from its family business foundation. It had gone public with a stock sale in the mid-1960s and later was sold to AMF, a bowling ball manufacturer. The 1970s was a dismal decade for American automakers and for Harley-Davidson. Japanese imports were popular while the quality of American-made cars and motorcycles was suspect in the minds of many consumers. Harley-Davidson's dealers complained that the new bikes they received sometimes needed a lot of repair work.

Harley-Davidson seemed lost in a corporate malaise while losing some of its loyal customers to the Honda Gold Wing. Behind the scenes, however, a member of the original families was preparing to help rescue the company. Willie G. Davidson, Walter's grandson, worked in the design division. He began designing new models based on Harley customers he met at rallies and swap meets.

In 1981, a group of Harley managers bought the unprofitable company from AMF for $81.5 million in a leveraged buyout. A recession hit the country during those early days of the Reagan administration, but Harley-Davidson was on its way back. The federal government finally gave the company some respite from the unfair trade tariffs, imposing a tariff on some Japanese motorcycles. Harley-Davidson began promoting itself as the only American motorcycle manufacturer, incorporating the U.S. flag into its advertising. Under the design leadership of Willie G. Davidson, the company revived some of the older designs that had built its popularity decades earlier. The company also overhauled itself on the inside, adopting quality control practices of its Japanese competitors.

Harley-Davidson's efforts won a new generation of white-collar customers and made its bikes chic among Hollywood celebrities. Annual rallies at Daytona Beach, Fla., and Sturgis, S.D. celebrated the motorcycle and its long history. New models were introduced during the 1990s that were applauded by the company's growing base of customers.

In 2003, Harley-Davidson celebrated the centennial of a modest company founded by Milwaukee brothers that evolved into an international corporation and an American icon.

OLE & BESS EVINRUDE

Ole Evinrude was wooing Bess Carey in 1904 and he wanted to buy her an ice cream cone. The problem was that Ole and Bess were on the wrong side of Wisconsin's Okauchee Lake and he would have to take a boat across the lake to buy the ice cream. Ole rowed across the lake and returned with the ice cream, which, under the blazing sun, had become a soupy mess.

The embarrassing experience could have inspired him to invent a portable refrigeration unit. Instead, he was inspired to fix up his rowboat with an outboard motor.

By 1907, Evinrude completed the design of his motor. It consisted of a horizontal cylinder, a vertical crankshaft, and directional gears submerged in the lower section. The design remains standard today. The design led to the founding of Evinrude Motor, the Outboard Marine Corporation and later to the Lawn-Boy lawn mower.

But Evinrude wasn't the only person who wanted to manufacture boat engines. The Johnson brothers — Lou, Harry, and Clarence

Wisconsin State Journal archives

Ole Evinrude

— of Terre Haute, Indiana, also built a marine engine and they later founded the Johnson Motor Wheel Company. Another competitor produced a detachable rowboat motor called the Waterman Porto Motor, which made such a racket that the marketing slogan was "Don't be afraid of it." The superior engineering of Evinrude's motor soon drove the Waterman model off the market.

Evinrude planned just to build motors for the local market. But as word spread about his motors, orders flooded in. In 1910, Bess Carey, who overlooked the soupy ice cream and by then was Mrs. Evinrude, wrote the company's first advertisement: "Don't row. Use the Evinrude detachable boat motor." She managed the business while he supervised the shop. A friend, C.J. Meyer, contributed a thousand dollars and became a partner in the fledgling company.

When Bess Evinrude launched her national advertising campaign, Ole had to hire a hundred men to built motors. The company was strapped for cash as it struggled to meet the national demand. Mrs. Evinrude later explained how the company survived.

"By turning materials into finished motors and selling the motors for cash before the bill on the materials was due, he made a hundred dollars do the work of a thousand in the ordinary plant. And we worked! There wasn't a night that we closed our eyes before twelve or one o'clock, and some nights it was two or three."

The national success spurred Bess Evinrude to look for orders on an international scale, partly to overcome the seasonal nature of the business. She contacted export houses and got one firm to stock the motors after the Danish manager of the company's Scandinavian department decided he could sell motors to Scandinavian fishermen. They started with two motors and later ordered thousands.

By the end of its third year in business, the Evinrude Company had three hundred employees and a new factory. But the stress and long hours were catching up with Bess Evinrude and her health was deteriorating. Ole Evinrude decided to sell his share in the business to Meyer. The deal included a provision that Ole Evinrude was prohibited from getting involved in the outboard motor business for five years.

The Evinrudes toured the country with a bed in the back seat of their car and sailed down the Mississippi River in a boat with an engine designed by Ole. But Ole couldn't keep from tink-

ering and by 1917 he was designing another
motor. The ELTO (Evinrude Light Twin Outboard) was a two-cylinder engine and the first
of its kind. It weighed twenty-seven pounds less
than Evinrude's earlier one-cylinder model but
offered three horsepower compared to two for
the single-cylinder engine. The secret was that
Ole had substituted aluminum parts for brass
and iron.

Ole Evinrude took his new motor to Milwaukee, but Meyer wasn't interested in it because
the company was doing well enough with the
earlier design. So Evinrude founded the Elto
Outboard Motor Company to manufacture the
new engine. The Evinrude team was in business
again with Bess serving as the company's secretary and treasurer. They competed against
their former company. Clinging to old technology, the old company began a gradual decline
while Evinrude's new company became successful.

In 1926, Evinrude unveiled the Super Elto
Twin that he hoped would capture the market.
He didn't count on a newly designed motor from
the Johnson Motor Company that weighed
nearly a hundred pounds but could power a
boat at sixteen miles per hour. Evinrude's motors could manage only ten miles per hour. Consumers chose speed over weight and Evinrude's
sales suffered.

As a recreational, non-essential product, outboard motors didn't sell well during the early years of the Depression. The selling price of Evinrude's motors dropped from $115 to $34.50, but buyers began to value smaller and lighter engines.

In 1929, the original Evinrude Company merged with Elto and the Lockwood Motor Company of Jackson, Michigan, to form the Outboard Motors Corporation. Evinrude was president of the new corporation and Stephen Briggs was chairman. Briggs and Harry Stratton were pioneers in developing small gasoline engines.

The Johnson Motor Company went into receivership in 1932, tried to manufacture refrigerator compressors and then was acquired by Ralph Evinrude, Ole's son, in 1935. Johnson merged with the Outboard Motors Corporation the following year and the new company eventually manufactured sixty percent of all outboard motors sold.

Bess Evinrude died in 1933 and Ole died a little over a year later on July 12, 1934. The same year, the company manufactured the original Lawn-Boy power lawn mower. By the late 1930s, Ralph Evinrude was president and a major stockholder of a corporation that employed two thousand men.

HARRY HOUDINI

Long after his death, Harry Houdini still is known for his breathtaking escapes and magic tricks. He was a master showman and reinvented magic.

Houdini was born in 1874 as Ehrich Weiss in Budapest, Hungary, to Rabbi and Mrs. Mayer D. Weiss. While Ehrich still was a baby, the family came to the United States, settling in Appleton, Wisconsin. After he became famous, Houdini often claimed he was born in Appleton and sources list his birth date as either March 24 or April 6. A 1969 book titled *Houdini: The Untold Story* says his parents were named Samuel and Caecilia Weisz.

At age 5, he saw a traveling circus that came to town and happened to see how one of the magician's tricks was done. Young Ehrich was hooked on magic and three years later joined a traveling circus, performing as "Eric, Prince of the Air." As a contortionist, acrobat and magician, he toured the state with several traveling circuses. While performing as a contortionist, he first attempted one of the escapes that later would become his trademark.

Houdini gained worldwide fame based on the

Wisconsin State Journal archives

Harry Houdini prepares another escape.

fact that people simply couldn't believe that a man could survive the torture chambers or escape the handcuffs he fastened on himself. He also escaped from caskets, ropes, mail bags, and once from the belly of a whale in a re-enactment of the biblical story of Jonah.

Some people thought Houdini must have supernatural powers to effect his escapes, but he often tried to expose spiritualists as fakes. The escape from the whale prompted him to make an after-death pact with theater manager John Royal.

"All right, if I can come back to earth, I certainly will be able to do it a month or six weeks after death," Houdini said. "If I should die first, I'll come back and say to you, 'Do you remember the fish you put me in in Boston?' Should you come back first, you might ask me if I remember how long the fish had been dead."

In his later years, Houdini appeared in movies and wrote books on the art of magic. He offered a $10,000 reward to anyone who could develop a stunt that he couldn't perform after watching it three times. The reward never was collected.

Houdini died on Halloween in 1926 at age 52 after an attack of acute appendicitis evolved into peritonitis. Houdini pledged to "get out of this the way I get out of everything else." But after a week of battling the illness, he admitted to his brother, Theodore Hardine, that he was "all through fighting." Moments before his death, he uttered an incoherent reference to Col. Bob Ingersoll, a famous agnostic of the time.

Similar to the Elvis Presley legend, Houdini's fame persisted long after his death. Attempts were made to contact Houdini beyond the grave. In 1976, Warren Freiberg, a radio talk show host and self-proclaimed psychic detective, claimed to raise Houdini's spirit by putting his wife in a trance during his Chicago-area radio show. An official Houdini seance was held every year since his death. These efforts were made even though in life Houdini often debunked claims that people could communicate with the dead.

Houdini's widow was among the participants in a 1943 seance.

"Harry could escape from anything on earth," she said. "If he can't slip through a message to me from heaven then the whole deal is off." After attending a decade of seances, Beatrice Houdini said shortly before her death at age 67 that she was convinced that the dead cannot communicate with the living.

But in other ways Houdini did speak after his death.

Voice recordings made by Houdini on a primitive three-inch cylinder were found among the effects of a famous musician, John Mulholland, who died in 1970. In the recordings, Houdini described his water torture cell and offered a thousand dollars to anyone who could prove that he could get air while locked in the cell.

In 1974, a letter Houdini wrote in 1926 about how to survive for 91 minutes in a sealed iron coffin with only five minutes of air was discovered. He had written the letter shortly before his death to a mining consultant hoping his technique would help trapped miners. He concluded that fear rather than actual lack of air killed many trapped miners.

PLEASANT ROWLAND

Two experiences came together to inspire Pleasant Rowland to start the Pleasant Company, which became the manufacturer of American Girl books and dolls.

She visited Colonial Williamsburg in 1984 and enjoyed revisiting the places of George Washington and Patrick Henry.

"I remember sitting on a bench in the shade, reflecting on what a poor job schools do of teaching history and how sad it was that more kids couldn't visit this fabulous classroom of living history," she said in an article published in *Fortune Small Business Magazine*.

When Christmas came, she wanted to buy dolls for her two nieces aged 8 and 10. But the Cabbage Patch Kids were all the rage and her only other alternative was Barbie. It was then that she had an idea: She would create a series of books about girls growing up in various historic periods with dolls for the characters that the girls could use to play out the stories.

Rowland was fortunate that she had saved $1.2 million to start her business. She was 45

Wisconsin State Journal archives

Pleasant Rowland, center, with doll manufacturer Marian Gotz, left, of Rodental, West Germany, and rattan manufacturer Elly Tiuti of China.

years old, and a large share of the money came from royalties she had earned as the author of reading textbooks. She also had worked as an elementary school teacher, magazine publisher, and TV news anchor.

She wrote a detailed business plan and began searching for a doll to use as a model for her character dolls. A friend found one relegated to a dusty basement storeroom at Marshall Field's in Chicago due to the doll's crossed eyes. When she undressed the doll, she found the name of the manufacturer sewn into the underpants: Goetz Puppenfabrik of Rodental, West Germany. The books were produced in America, and miniature accessories were made in China.

Although the Pleasant Company was headed for a 1986 launch, some people told Rowland that her idea would never work. She held a focus group of mothers who also didn't think much of the concept until they were able to see a doll and sample book. At that point, they seemed to like it.

Rowland knew that her books and dolls weren't the kind of items that would sell well at Toys 'R' Us or in flashy television commercials. She decided that direct mail would be her best marketing method so she mailed a half million catalogs the first Christmas. The idea and the marketing worked. During the last four months of 1986, her company's net sales were

$1.7 million.

"That first Christmas we cobbled together packing stations out of plywood and old doors," she told *Fortune Small Business.* "We were in a broken down warehouse with one freight elevator. Workers wore mittens because there was no reliable heat."

When sales grew to $7.6 million the second year, Rowland was able to afford to build a new warehouse and company headquarters outside of Middleton, Wisconsin.

Rowland aimed first for an upscale market, telling employees to "picture mothers who drive Volvos," according to a 1996 article in the *Wisconsin State Journal.* Her customers included director Ron Howard, actress Jessica Lange and Ivana Trump.

By 1998, the company's annual sales had reached about $300 million. Rowland opened American Girl Place in Chicago, which featured a musical play called *The American Girls Revue.* That year she also sold her company for $700 million to Mattel. She felt a connection to Jill Barad, then Mattel's chief executive officer, who had built the company's Barbie sales from $200 million to $2 billion.

In the *Fortune Small Business* article, Rowland said she was surprised that she felt no emotion when signing the papers for the sale.

"It was then I realized that I had never felt I owned American Girl," she said. "I had been its steward and I had given it my very best during the prime of my career. It was time for someone else to take care of it."

Rowland's husband, Jerry Frautschi, who owned a printing company in Madison, Wisconsin, used a large part of their fortune to create a cultural center west of Madison's Capitol Square. He donated $200 million to create the Overture Foundation, which rebuilt a block of prime business property along State Street.

American Girl books and dolls targeted preteen girls as customers for the first time. The Pleasant Company also provided a more wholesome and educational alternative to the Barbie phenomenon.

INCREASE LAPHAM

Colorful weather maps in newspapers and on television can trace their existence to Increase A. Lapham. He was Wisconsin's first scientist who pioneered weather forecasting and promoted establishing a national weather bureau.

Lapham's scientific interests were wide-ranging, and he also is known for cataloguing the state's Indian mounds and debunking theories that mounds were built by a "lost race" instead of ancestors of Midwestern tribes. He was an author, engineer, botanist, geologist, meteorologist, conchologist, and archeologist. He also advocated conservation of Wisconsin's forest and was a founder of Milwaukee Female Seminary, the first Milwaukee high school which became Milwaukee-Downer College.

He also served on the Milwaukee School Board, helped found the Milwaukee Public Library, and served as president for a decade of the State Historical Society of Wisconsin. He drew some of the earliest maps of the state.

He was born in New York in 1811 and worked with his father, who was a contracting engineer, on projects in Ohio and Kentucky as well as on

Wisconsin State Journal archives

Increase Lapham

the Erie Canal. He came to Milwaukee in 1836 at age 25 to work on a project known as the Milwaukee and Rock River Canal and to help Byron Kilbourn survey the then-village of Milwaukee.

By 1842, Lapham's strong interest in the natural environment led him to develop an interest in weather forecasting. Many people thought the idea was ridiculous, but he began a long campaign for a national weather bureau, which the federal government initiated in 1869.

He had seen Indian mounds in Ohio, and Lapham's interest in them continued when he moved to Wisconsin. At the time, they weren't valued or protected from farmers who often plowed them under for crops.

Lapham's books included *A Catalogue of Plants and Shells Found in the Vicinity of Milwaukee,* published in 1836; *A Geographical and Topographical Description of Wisconsin,* published in 1844 and viewed as the first scientific work of Wisconsin; and *The Antiquities of Wisconsin, as Surveyed and Described,* published by the Smithsonian Institution in 1855 and republished by the University of Wisconsin Press in 2001.

Like weather forecasting, archeology wasn't a reputable science when Lapham began studying the remains left by earlier civilizations.

In his later years, Lapham turned his interests to Wisconsin's extensive forests, studying them and working to preserve them.

In the 1860s, he was called upon to write the federal law that established a national weather bureau, later to become the U.S. Weather Service. He was appointed Wisconsin's chief geologist in 1873 and served two years in the post.

In 1875, he was asked by the Smithsonian Institution to build an effigy mound model for display the following year at the Philadelphia Centennial. Lapham completed one model and was asked to do another. But he died Sept. 14, 1875, in his rowboat on Oconomowoc Lake.

Lapham's scientific interests were so wide-ranging that when he was asked late in life what common thread they might have, he replied: "I am studying Wisconsin."

STEPHEN BABCOCK

Before Stephen Moulton Babcock came on the scene in 1890, dairy farming in Wisconsin was a risky proposition. Milk was sold by creameries with no quality standards. Farmers complained about the prices they got for their milk, and many began thinning it with water to get a better price. The only butterfat test then in existence took two days and was rarely used.

Stephen Babcock

Babcock, who already was known for his scientific expertise, was asked by Dean Henry, director of the Wisconsin Experiment Station, to devise a simple test for butterfat in milk. After several months, Babcock came up with a test that revolutionized the dairy industry because for the first time it allowed milk to be sold based on its quality. His test also helped spawn the entire cheese industry.

The test, which took about five minutes, used sulfuric acid to dissolve all nonfat solids in milk, heat melted the fat, and a centrifuge spinning the bottle forced it to rise to the surface. That meant the fat could be measured in the necks of the bottles.

Babcock declined millions of dollars in royalties for his simple test, which soon was adopted worldwide and wasn't changed during his lifetime.

"I fully expected that the markings on the tubes would be changed," he said shortly before his death in 1931.

Babcock was born in Bridgewater, N.Y. in 1843 and graduated from Tufts College in 1866. He taught at Cornell University and earned a doctor's degree at the University of Goettingen in Germany. When he returned to the United States, he resumed his position as a chemistry professor at Cornell.

When he came to Madison, Wisconsin, in 1888, the enrollment at the University of Wisconsin's College of Agriculture totalled two students and Babcock was one of two professors. He devised the butterfat test in a barn on the campus.

Babcock's research also led to the recognition of the nutritional values in food and the discovery of vitamins. He was one of the first to realize that food contained then mysterious elements that controlled growth, energy, health and life itself.

Besides his scientific pursuits, Babcock, a lifelong bachelor, enjoyed baseball and gardening.

School children often visited his home and he would talk to them about his inventions and a cow named Sylvia who was involved in developing the butterfat test. He had been blind in one eye since childhood, and bought his first car at age 76.

"I drove up town and tried to run over a streetcar," he told his housekeeper. "But I found it couldn't be done and I've never tried it again."

More than a decade after his death, Wisconsin officials and others marked the centennial in 1943 of Babcock's birth. John H. Kraft, president of the Kraft Cheese Co., presented a $7,000 check to the university to further dairy research.

"Had I visited the university and seen the concrete evidence of the work stemming from his discoveries before the amount of the grant was decided upon, the amount would have been larger," Kraft said.

HARRY ARMENIUS MILLER

While the Harleys and Davidsons popularized the motorcycle, they didn't invent it. That distinction belongs to Harry Armenius Miller of Menomonie, who attached a one-cylinder engine to a bicycle. He also built the first outboard boat engine, long before Ole Evinrude made a fortune from manufacturing them.

Miller was born December 9, 1876, to Jacob Miller and Martha (Tuttle) Miller. Harry Miller's parents were German immigrants and more than anything they wanted him to get a good education. But young Harry had other ideas. He had mechanical skills, and his father was upset that he dropped out of high school at 15 to work in the Knapp, Stout & Co. machine shop.

Two years later, young Harry left Menomonie for Salt Lake City. He returned to his hometown, but then went to Los Angeles, where he found a job in a bicycle shop and met and married Edna Inez Lewis. The couple ultimately returned to Menomonie, where Miller resumed his job in the machine shop.

A steep hill between Miller's home and the machine shop may have inspired him to mount a small engine on a bicycle. In the mid-1880s, Miller put a four-cylinder engine on a rowboat, inventing the first outboard motor.

Harry and Edna Miller returned to Los Angeles in 1897 and he opened a machine shop, building his first automobile in 1905. He also designed engines for inboard racing boats and airplanes.

Besides being the inventor of the motorcycle and the outboard motor, Harry Miller designed racing car engines. In 1916, Barney Oldfield, then a top racing driver, discovered Miller, and they collaborated on the Golden Submarine, a car that brought Miller wide recognition.

In 1920, he built a 183-cubic-inch engine that helped Jimmy Murphy win the Indianapolis 500 two years later. During the 1920s, cars designed by Miller won 73 of the 92 major races in the nation. At the 1929 Indianapolis 500, 27 of the 33 cars were powered by Miller's engines.

Later, Miller helped develop the Tucker automobile during the 1930s. He died of cancer in 1943.

In 1999, the Miller-Offenhauser Historical Society was formed to preserve and publicize the racing designs of Harry Miller, Fred Offenhauser, and Leo Goosen. The society's goal is to collect and memorialize the products created by these men from about 1917 to 1945.

JOHN WESLEY CARHART

Long before Henry Ford popularized the automobile, John Wesley Carhart invented what may have been the first auto by putting a two-cylinder steam engine on a buggy. He called his invention the Spark.

But Carhart, then living in Racine, was forced to dismantle the machine after it killed a horse that was startled by the noise.

Carhart was born June 26, 1834, in Albany County, New York.

Like other inventors, John Carhart liked to tinker, and he once made a miniature steam-powered yacht that he sailed on the Hudson River. He was ordained a Methodist minister at 17 and earned his doctor of divinity degree in 1861 from Union Seminary in Charlottesville, New York. He married Theresa Mumford and they had eight children.

Carhart served as pastor at churches in New York, Massachusetts, and Vermont. He published two books of poems called *Sunny Hours* (1859) and *Poets and Poetry of the Hebrews* (1865).

Preaching and writing didn't stop Carhart from pursing his inventions. He built an oscillating valve for steam engines that he patented and sold for several thousand dollars. But his steam-powered buggy didn't come about until he was transferred to Racine in 1871.

The buggy made such a racket that it terrified bystanders and horses.

After Carhart moved to a church in Oshkosh in 1874, his children began publishing a weekly newsletter, the Early Dawn, that was printed on a steam-powered press he had set up in the church basement.

By 1880, Carhart had accomplished enough to publish an autobiography titled *Four Years on Wheels.* He renewed an earlier interest in medicine, becoming a doctor in 1883 and practicing in Oshkosh.

In 1885, Carhart and his family followed their son Ed to Texas, settling first in Clarendon and then moving to Lampasas. He started a weekly newspaper before setting up a medical practice. He developed a reputation as an outstanding skin and nerve specialist in La Grange, Austin, and San Antonio.

Carhart also became an advocate for better sanitation practices and served as an assistant secretary general of the Pan American Medical Congress in 1893 in Washington, D.C.

Carhart published a novel, *Norma Trist,* in 1895 that was one of the first novels to explore homosexuality. Although it was moralistic, Carhart's book outraged critics, and he was arrested for sending obscene literature through the mails, although the case eventually was dismissed. In 1899, Carhart published *Under Palmetto and Pine,* a book about the struggles of African Americans in Texas with poverty and racial discrimination.

The magazine *Horseless Age* recognized Carhart as "the father of the automobile" in 1903, and he was invited two years later to Paris as a guest of the French government at the International Automobile Exposition, where he received a cash award and certificate to honor his invention.

Carhart — author, doctor, inventor, minister, newspaperman — died December 21, 1914 at San Antonio and was buried in Austin.

CHRISTOPHER LATHAM SHOLES

As with many inventors, Christopher Latham Sholes of Milwaukee didn't set out to invent what he eventually invented — the first practical typewriter. He was working on a machine that could number book pages in sequence when a chance remark by one of his coworkers caused him to change direction.

"Why cannot such a machine be made that will write letters and words and not figures only?" asked Carlos Glidden, who was developing a mechanical "spader" to take the place of a plow. Some accounts also say that the idea for the machine was suggested to Sholes by General William La Due.

A year later, Glidden came across an article in *Scientific American* magazine that described a machine called a "pterotype" invented by John Pratt, that apparently did just as Glidden had suggested. Glidden, Sholes and Samuel Soule set to work. By the fall of 1867, they had fashioned a crude version of the machine on which they wrote several letters to friends, including James

Wisconsin State Journal archives

Christopher Latham Sholes and his typewriter.

Densmore of Meadville, Pa. Densmore immediately envisioned the machine's value and offered to finance its development.

The first machine was rigged up with a single telegraph key and printed the single letter "w" through carbon paper. When a model was complete that typed all of the alphabet, the first sentence Sholes typed was: "Now is the time for all good men to come to the aid of their party." The sentence came from a newspaper editorial of the time.

About 30 models were produced and tested. Many of the prototypes were built by Mathias Schwallbach, a clock maker skilled at building models. Schwallbach sometimes offered suggestions to improve the machine.

It took until 1873 to get a machine suitable for manufacture. James Ogilvie Clephane, a shorthand reporter in Washington, D.C., tested one model after another for Sholes and his colleagues. Clephane later was closely identified with Ottmer Mergenthaler, inventor of the Linotype.

Densmore contacted with E. Remington & Son, a gun manufacturer, and persuaded them to manufacture the finished machine. Early models resembled a sewing machine, which Remington also built, complete with a foot pedal. Some earlier typewriter models were as large as pianos and used the same design concept. He also developed the "qwerty" keyboard, an unusual way of arranging letters still used on computer keyboards today.

Sholes sold his share in the typewriter for $12,000, according to some accounts. When the machine hit the market, over a hundred people claimed to be the inventor. Sholes didn't dispute any of them, but he held the patents and the manufacturing contract.

Sholes was born in 1819 in Pennsylvania, where he was an apprentice printer. At age 14, he came to Wisconsin to join his brother in Green Bay. As a young man, he got involved in politics, serving as a senator and in the state Assembly. In the 1830s, he became editor of *The Wisconsin Inquirer,* a newspaper published in Madison. He lived in Kenosha for several years, editing a newspaper called the *Telegraph Courier* and serving as postmaster. He moved to Milwaukee in 1860, where he served as postmaster and later as commissioner of public works and customs collector. He also worked as editor of the *Milwaukee Sentinel* and the *Milwaukee News.*

Although he was confined to bed during the final years of his life, Sholes continued to perfect the typewriter. He died Feb. 17, 1890, at age 71.

According to a 1927 article in *Wisconsin Magazine,* Sholes believed his invention helped women to more easily make a living. He wasn't aware of the full magnitude of his work.

By the time a plaque was dedicated to him in Milwaukee in 1924 to mark the 50th anniversary of the invention, typing pools clattered in offices throughout the nation. Sholes also was honored by a mural in the Milwaukee County Courthouse.

Word of the anniversary honors bestowed upon Sholes reached Alabama, where the State Federation of Women's Clubs launched a campaign to honor John Pratt of Center, Ala., as the true inventor of the typewriter. Pratt exhibited his device at the Royal Society of Arts and Science in London. Unlike Sholes, he never got it into mass production beyond about thirty machines he sold in England. Pratt went to London with the invention because he believed a Southerner wouldn't get recognition for it so soon after the Civil War.

Since Sholes popularized the typewriter, it evolved into quieter electric models and today into computer keyboards. The function, however, remains the same.

WALTER KOHLER

Walter J. Kohler took a plumbing business started by his father and grew it into a worldwide corporation. He also founded the Wisconsin community of Kohler as a place for his employees to live.

During the industrial revolution of the late 1800s and early 1900s, many corporations built "company towns" of housing for their workers. Many of these towns bound the workers more tightly to their employers because they usually rented their homes. In Walter Kohler's town, workers owned their homes and governed themselves.

The Kohler company was founded by Walter Kohler's father, John Michael Kohler. John Michael's father was a hotel proprietor in Austria, and the family came to America in 1850 when John Michael was 8, settling on a farm near St. Paul, Minnesota.

When he was 16, John Michael sought work in Chicago, where he met and married Lilly Vollrath, the daughter of Prussian immigrant Jacob J. Vollrath. When the great Chicago fire burned out the Kohlers, the family moved to Sheboygan, and John Michael, then 27, founded the Kohler Co. to manufacture farm implements.

Among the implements manufactured by the Kohler Co. were pig troughs, which John Michael decided could make fine bathtubs. That idea propelled the company into the manufacture of enameled bathroom fixtures. John Michael Kohler had the idea that bathrooms, then fairly new with the advent of indoor plumbing, some-day would evolve into a luxurious part of the home.

Walter Kohler was born in Sheboygan and at-tended elementary and high school before go-ing to work at age 15 in his father's business. Young Walter started in the enameling depart-ment earning $7.50 a week for six work days.

"When I was a boy, we broke up our pig iron by hand," he later recalled. "It is all done by machinery now, but the old method built some good strong hands on the young fellows."

On his 18th birthday, Walter Kohler was named foreman with a raise to $12 a week. He worked 12 hours a day and once worked 60 con-secutive hours in sweltering July heat when the plant had a breakdown. As foreman, Kohler re-duced the length of shifts for enameling furnace workers to eight hours.

In 1899, the Kohler Co. moved from Sheboygan to the future site of the village of Kohler, and a year later Walter Kohler married Charlotte Schroeder, a Kenosha teacher. John Michael Kohler died two days later, leaving three

Wisconsin State Journal archives

Walter Kohler

sons to carry on his work. But it was Walter Kohler who had the strongest impact on the company and the state.

More than a decade after his father's death, Walter Kohler founded the village of Kohler as a home for his employees and for his company, a community of vine-covered homes, well-kept lawns, hedges, and flower pots. A dance hall was turned into a meeting place for the Kohler Village Women's Club.

Kohler began providing worker's compensation for laid-off workers two years before it was required by state law. He offered insurance and health benefits in which the company paid half the cost. He built a place with coffee and cigarettes for employees to take their breaks and a shelter for people who came to fill out job applications.

"We are trying to make the Kohler Co. the best possible place to work and our community the best place to live," he said.

Walter Kohler got involved in politics when a Wisconsin delegate to the 1928 Republican National Convention died and Kohler was chosen to replace him. Five days after the national convention, Kohler was endorsed as a candidate for Wisconsin governor. In the primary, he defeated Republican incumbent Gov. Fred Zimmerman and Joseph Beck, a candidate backed by the La Follette progressives. Stung by the defeat of their candidate, progressives threw their support to then-Madison Mayor Albert Schmedeman, but Kohler prevailed.

The progressives charged that Kohler violated the state corrupt practices act, claiming that $100,000 was spent on his behalf even though Kohler himself abided by the campaign spending limit of $4,000. A Sheboygan County judge dismissed the claim, but that decision was reversed by the Wisconsin Supreme Court. Kohler ultimately was vindicated after a circuit court trial.

During his two-year term, Kohler balanced the state budget and joined with the Legislature to prohibit "yellow dog" contracts, which required employees to pledge not to join a union as a condition of employment. He failed to get legislative backing for an ambitious plan to link the state with hard-surface highways. He also failed in an effort to reduce funding for the University of Wisconsin.

Progressives rallied from the 1928 defeat and two years later Phil La Follette outpolled Kohler in the Republican primary and went on to become governor. In 1932, some Republicans pressured Kohler to run against La Follette and he finally consented. Although he beat La Follette in the primary, the strength of President Franklin Roosevelt's landslide election swept Kohler's old Democratic opponent, Schmedeman, into office. The *Wisconsin State Journal* suggested in an editorial that Kohler would be an ideal Republican presidential candidate in 1936.

But Kohler retired from politics after the hard-fought 1932 election. Despite his labor reforms, Kohler was too conservative for La Follette progressives, aligning himself with President Herbert Hoover and opposing Roosevelt's New Deal reforms. In 1934, he wrote in a newspaper column that increasing cheese consumption was a way to improve Wisconsin's economy, noting that Americans eat less than half as much cheese as Europeans each year.

The idyllic image of Kohler Co. and its employee village was shattered in 1934, when the American Federation of Labor tried to organize employees and replace the company union. The effort split the work force into opposing factions. On July 27 that year, a violent confrontation between those factions left two men dead and 21 wounded by sheriff's deputies. Gov. Albert Schmedeman sent in the state militia to restore order and Kohler blamed "outside agitators" and Communists for the violence.

In his later years, Kohler spent much of his time at his home called River Bend in his namesake village. When he traveled, he often flew from the Kohler airport in a monoplane similar to Charles Lindbergh's "Spirit of Saint Louis." Kohler's plane was called "Village of Kohler."

On April 21, 1940, Kohler's wife became concerned when he failed to appear for his morning horse ride. She found him huddled under a blanket on the floor of the bedroom. Then 65, he died of thrombosis or a blood clot in the heart.

The company Kohler propelled to international prominence still produces quality plumbing fixtures and the village of Kohler still prospers.

AL RINGLING

Perhaps it was a stern family upbringing that caused Al Ringling and his four brothers to become fascinated with the circus. Their parents, August and Salome Ruengling, were strict Lutherans who believed that circus people were evil.

Like a roving circus troupe, the Ruengling family moved around quite often. August and Salome were married in Milwaukee. They lived in Chicago, then moved to Baraboo. During the Civil War, the family fared well, but afterward an economic crash devastated the family's finances and they moved to McGregor, Iowa.

McGregor is across the Mississippi River from Prairie du Chien, Wisconsin, where entertainment troupes used to arrive by riverboat during the post-Civil War years. Ringling and his brothers went to the dock to watch the arrival of the Dan Rice show. The boys watched open-mouthed as an elephant lumbered down the gangplank. Later, they were thrilled when the circus strongman visited their father's harness shop to have a strap repaired.

The boys were hooked. After moving back to Baraboo, Al practiced juggling plates while John learned to sing and dance. Alf T. and Charles learned to play musical instruments while Otto organized pin games for neighborhood children.

After several neighborhood performances in Baraboo, the Ringling boys decided to take their show on the road. Al took $3.75 from his savings to print handbills, and a performance was scheduled in a nearby village. The show drew an audience of 47 people, and the boys began planning a more ambitious tour. In November 1882, they traveled to Mazomanie for their first road performance as the Ringling Brothers Classic and Comic Concert Company.

"I was never again to be uneasy as on this first appearance," Al said much later. "We had gone far enough from Baraboo that no one

Wisconsin State Journal archives

Al Ringling performs with a troupe of trained dogs in this 1887 photo.

would know us. On the afternoon of the day of the show, we paraded the streets. There were 59 paid admissions to the performance, enough to meet the hotel expenses and have a little left over."

With Al in blackface, Alf T. playing the drums and Charles on trombone, the brothers toured Iowa and the Dakotas before returning to Wisconsin to close the season at Oregon in Dane County. They had earned $300, more than their father's harness shop income, and also had bought new clothes.

With the money from their successful tour, the brothers bought a monkey and a hyena and started their first real circus in 1884. The boys toured most of the year and established their winter quarters in their hometown of Baraboo. Their most dramatic act featured Alf as "the man with the iron jaw," balancing a farm plow on his chin. The proceeds went to buying more animals and making their show stronger and better. By 1888, the circus had two elephants and a band. Al served as ringmaster before performing in a cannonball act. The circus now was grossing $15,000 a year.

Four years later, the Ringling Brothers Circus rolled into Milwaukee in eighteen rail cars. Besides the usual elephants and monkeys, the show also featured two hippopotamuses. In Milwaukee, the Ringling circus went head-to-head against a larger rival: Barnum & Bailey. The

Ringling show was standing-room-only.

In 1908, the Ringling Brothers bought out Barnum & Bailey for $410,000. The Barnum & Bailey circus was operated independently for a decade until 1919, when the two circuses merged into what was dubbed the Greatest Show on Earth.

A key to the phenomenal success of the Ringling Brothers was their creativity. As the circus grew, the humble little musical band was supplanted by pageants highlighting biblical, historical or mythical figures such as Solomon, Joan of Arc or Cinderella.

Despite their worldwide travels and successful barnstorming tours throughout the nation, Baraboo remained home to the Ringlings. Al Ringling built a $135,000 house in 1909 and opened the Al Ringling Theatre in 1915.

But the circus moved out of Baraboo after the merger with Barnum & Bailey, and the combined circus wintered in Connecticut.

The giant circus returned to Baraboo in 1933 for a special tribute. Elephants appeared in gold paint and more than 40,000 people gathered for the first Ringling hometown performance in over 35 years. By this time, John was the last living brother. Otto had died in 1911, Al in 1916, Alf in 1919, and Charles in 1926.

Besides their worldwide contribution to the circus, the Ringling brothers also were the first of Wisconsin's master showmen. In later years, this tradition was carried on by Tommy Bartlett of Wisconsin Dells and Alex Jordan, developer of the House on the Rock.

THE JOHNSON
FAMILY

The Racine-based company known as Johnson's Wax and later as the S.C. Johnson Company traces its beginnings to 1886, when Samuel Curtis Johnson bought a parquet flooring business from the Racine Hardware Company. His profits the first year were $268.27.

Over the next century, the company would evolve into a $7 billion business and become one of the world's leading producers of household products. Unlike other mega-corporations, however, S.C. Johnson has at least two unusual aspects. The first is that it has been family owned throughout its history, as administrative control was passed from generation to generation. Although that was a common company structure in the early 20th Century, it is uncommon in these days of corporate buyouts and mergers. A second unusual aspect of the company is that it has kept its headquarters in Racine since the beginning instead of seeking cheaper labor elsewhere.

Within two years of buying the flooring busi-

Wisconsin State Journal archives

The first Samuel Curtis Johnson, at left, and his grandson, Samuel Johnson.

ness, Samuel Johnson had started producing Johnson's Prepared Wax and bought his first national advertising. In 1914, the company went international, setting up a subsidiary in Britain. Subsidiaries in Canada and Australia were established a few years later.

Throughout its history, the company has been sensitive to the value of its employees. Employees got paid vacations beginning in 1900 and a profit-sharing plan was established in 1917. A 40-hour work week was implemented in 1926 and in 1934, during the Depression, the company set up a pension plan. A child care center for children of employees opened in 1985, and during the 1990s the company repeatedly was named among the best hundred companies in America for working mothers.

The Depression years saw sales drop from $5 million to $3 million over a four-year period.

Although the company managed to survive the Depression, a turning point came in the mid-1930s that catapulted the business to even greater heights. H.F. Johnson Jr., who then headed the family business, had heard about the marvelous qualities of carnauba wax and that it was superior to waxes being made in the United States. He led the first expedition to South America in 1935 to study the carnauba palm, and the result was that the Johnson company soon began offering this special kind of wax. Carnauba wax, which coats the fronds of the palm, is the hardest natural wax in the world. H.F. Johnson Jr. made the trip in an amphibious plane and bought land to establish a plantation for growing the palm. He wrote a self-published book about the trip.

The company was so successful by the end of the 1930s that the Johnsons decided to build a new administration center. They hired renowned architect Frank Lloyd Wright to design it. When the center opened in 1939, it featured an area known as the Great Workroom, which was viewed as a center of creativity. Wright also designed the company's Research Tower, which opened in 1950. Both buildings were placed on the National Register of Historic Places in 1976.

The Johnson company also demonstrated

environmental concerns long before the environmental movement came along. In 1955, the company introduced aqueous-based aerosol sprays to reduce the environmental impact and in 1975 the company removed fluorocarbons from its products worldwide. The U.S. would ban fluorocarbons three years later. During the 1990s, the company received environmental awards in the United States, Canada, Indonesia, and Argentina.

Samuel Johnson became company president in 1965 when his father, H.F. Johnson Jr., suffered a stroke. Samuel Johnson had recently returned from the Netherlands where he had been assigned to set up an aerosol plant. He had run into problems with the plant, which was built with too much capacity and had start-up difficulties. H.F. Johnson Jr. died in 1978.

During the 1990s, under the leadership of Samuel Johnson, the company expanded its product offerings, adding Windex, Draino, Vanish, Ziploc, Saran Wrap, Pledge, Glade, and Fantastik. Samuel Johnson developed an insecticide, the company's first non-wax product, which later included Raid and OFF! By early 2001, the company had 9,500 employees in sixty-five countries.

As the 21st Century dawned and Samuel Johnson prepared to turn over the company to yet another generation, he revealed glimpses into the family that had built the multi-billion-dollar company. In a December 2002 article in *USA Today*, he talked about his father's drive to succeed and mistreatment of his family.

"Whether he loved the company more than me is really an issue that I struggled with," Johnson told *USA Today*. "He was so passionate about the company, so driven by that, there wasn't a lot of passion left."

Samuel Johnson, the namesake of the company founder, also spoke about his struggles with alcoholism that led to a stay at the Mayo Clinic in 1993.

In 1998, Johnson re-enacted his father's trip to Brazil in search of carnauba wax with a replica of a twin-engine amphibian plane called the Spirit of Carnauba. He also paid $4.3 million to a filmmaker to make a movie about the trip, but later sued the filmmaker claiming the first cut of the movie was too somber.

Johnson said in December 2000 that one key to the family's success was continual evolution and change.

"People don't always appreciate change the same way I do or perhaps our family does," he said in an Associated Press story. "In many ways the status quo is an enemy of companies in the sense that they don't adapt to a changing world and they don't adapt to a changing consumer or adapt to changing opportunities."

TOMMY BARTLETT

At age 71, a white-bearded and white-haired Tommy Bartlett performed in the water skiing show he created more than three decades earlier in Wisconsin Dells. It was the first time he skied in his own show and about four thousand spectators were on hand to witness the event.

Bartlett was a perfectionist who often sat in the audience scribbling handwritten critiques of his performers, The success of his shows along with intensive marketing turned his business into one of the most popular tourist attractions in the upper Midwest. It also financed his worldwide travels and fishing expeditions.

His show business career began at age 13, when he went to work for WTMJ radio in Milwaukee. He moved to Chicago four years later, hosting several programs for WBBM including one called *Meet the Missus* aimed at housewives.

He worked as a Northwest Airlines pilot and flight instructor during World War II, then returned to Chicago where he hosted a show called *Welcome Travelers*. The program was successful for a while and moved to television and Bartlett also hosted *The Tommy Bartlett Show* on the ABC network and made appearances on Don McNeil's *Breakfast Club.*

But Bartlett decided to get into a different business. He created a water-skiing business called the "Tommy Bartlett Water Circus" and had six touring shows performing daredevil stunts in colorful costumes. Inspiration for the ski shows came from a radio program he did from a fair on Chicago's waterfront that featured a similar Florida touring company.

"It was insurance against unemployment," he told the *Wisconsin State Journal* in 1979. "I always believe an entertainer should quit while he's at the top. He shouldn't just disintegrate."

At the time of that interview, Bartlett had never water skied. His skills were in promotion and marketing. In 1953, he found a home for his shows in Wisconsin Dells. Although the business made him a multimillionaire, he continued to play an active role, performing even mundane tasks like ushering spectators to their seats.

"We cater to what the public likes," he told the *State Journal.* "Meat and potatoes entertainment. There's never been an off-color joke in twenty-eight years."

Bartlett's business grew during the postwar boom when cars offered young families more mobility than they'd ever known and he provided a place to go. His water shows became a Dells fixture and helped build the community as a vacation destination for families from Chi-

Wisconsin State Journal archives

Tommy Bartlett tries water skiing for the first time.

cago, the Twin Cities, and other parts of the Midwest. He frequently spent winters in Florida, where he also operated a petting zoo during the 1960s.

The business was not without some problems. In 1976, the state Justice Department filed a lawsuit challenging Bartlett's exclusive use of a portion of Lake Delton for his shows. It appeared that Bartlett had been putting on his shows with the good will of other lake users, and local officials hastily drafted an ordinance giving him exclusive use for three fifty-minute periods a day.

At that time, the annual budget for the shows was $800,000, and they drew several hundred thousand people each summer. Had the lawsuits succeeded, they may have closed down the business or at least forced Bartlett to find a new location. But Sauk County Circuit Judge Howard Latton ruled in favor of Bartlett and a U.S. District Court dismissed part of a similar lawsuit filed by five fishermen who were removed from the lake by police after they refused to move their boats for a performance. In 1979, a U.S. Court of Appeals upheld those decisions and the performances continued without interruption.

To his performers, Bartlett became a grandfatherly figure, but his private life largely was a mystery. He was a lifelong bachelor who once said his best friend was Blitzen, a large Doberman. Bartlett died after a Labor Day performance in 1998.

ALEX JORDAN JR.

Some people believe that Frank Lloyd Wright designed the House on the Rock near Spring Green. The error is understandable because of Wright's strong identification with Spring Green and the fact that the House on the Rock has some Wright-inspired features.

But it was Alex Jordan Jr., the eccentric son of a Madison, Wisconsin, architect, who built the House on the Rock into Wisconsin's most popular tourist attraction.

Jordan's father designed the Villa Maria, a dormitory for women on the University of Wisconsin campus. He was so proud of the design, according to one account, that he decided to show the blueprints to Wright, presumably to gain the famous architect's approval. He traveled to Spring Green with Sid Boyum, a noted sculptor and lifelong friend of the younger Jordan, for the meeting with Wright. After looking at the blueprints, the story goes, Wright walked to the window, turned to the elder Jordan and told him: "I wouldn't hire you to design a cheese crate or a chicken coop."

Boyum said Jordan was angry as they drove

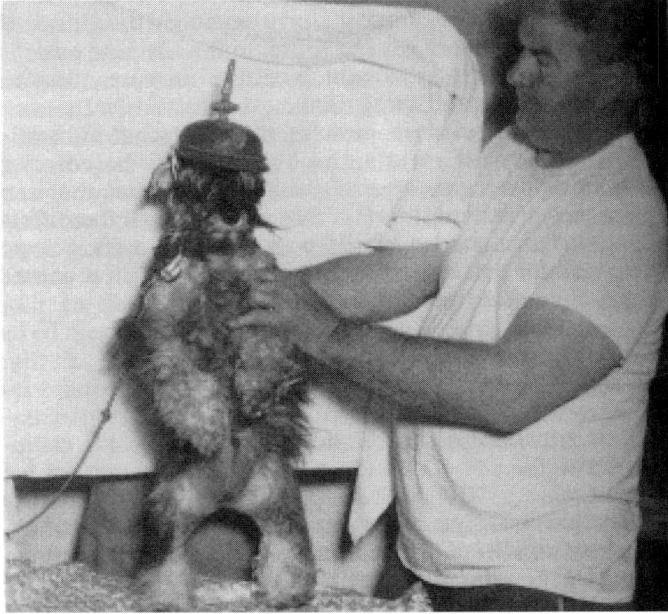

Photo courtesy of Sid Boyum

Alex Jordan Jr. tries to set up a photo of Tinkerbell, the family dog, wearing a German helmet he made.

south of Spring Green, stopping near the rock formation that later would be the site of the House on the Rock. Jordan vowed to get even with Wright by building a house on the rock.

The younger Jordan was getting into trouble in Madison at about the same time. Ever inventive, the younger Jordan had rigged a hole in a closet door of his girlfriend's apartment. He would hide in the closet while his girlfriend lured State Street businessmen to the apartment. Jordan would take pictures, then try to blackmail the businessmen.

It's unclear how many times the pair tried this Badger version of the old badger game, but both were arrested after Jordan demanded $35 or a job from one victim in exchange for the photos. After young Jordan served a brief jail term, the elder Jordan decided the house on the rock might be a good project to occupy his wayward son. He sent him out to Iowa County to supervise the project.

The younger Jordan was born March 3, 1914, and grew up in Madison. He was a spoiled child who often tested his parents' patience with his schemes. At St. Norbert College, he amazed his classmates with a magic trick in which he was able to guess what card was held face-out to his forehead. An accomplice hidden upstairs would signal the number and suit of the card to Jordan with an electrical gadget he had invented.

Junior, as he was known, warmed to the house on the rock project and soon a house was built accessible only by a rope ladder. The small house became a refuge for the younger Jordan as well as a party house. All he had to do was pull up the ladder and no one could come up. Windows on all sides provided a magnificent view of Wyoming Valley.

Curiosity caused many area residents to stop by the house on Sunday afternoon drives for a tour of the bizarre house. In 1959, Junior put up a bag, charged a quarter per person admission and opened the place to the public. For the younger Jordan, it was the beginning of his becoming one of Wisconsin's leading entrepreneurs.

Although the house was popular, Jordan worried that people wouldn't keep coming back unless he added something new every year. He started by putting goats up on a roof to graze, later adding an old-time street with lighted storefronts, a massive doll collection, and an orchestra of mannequin musicians.

Jordan took pride in picking up items at garage sales and flea markets and transforming them into pieces that appeared valuable. He made up fake histories for many of his artifacts. An example was the Tusk of Ranchipur, supposedly the lifelong work of a Punjab artisan that featured intricately carved figures on a long tusk. Actually, the tusk was made for Jordan by Dick Rahn of Mazomanie, and the name was a play on the words "Rahn is poor" — a shot at Jordan by Rahn for not paying him enough.

People kept coming back to the House on the Rock — a half million each year — and the original ticket price went up considerably. One of the final exhibits Jordan added before his death in 1989 was a gigantic battle scene in plastic of two sea monsters.

Shortly before his death, Jordan sold the attraction for $17 million to Art Donaldson of

Janesville, who operated a billboard company and had owned attractions at the Wisconsin Dells. Donaldson's business savvy kept the attraction popular into the 21st Century without the creativity of Alex Jordan Jr. He added a resort hotel.

FREEMAN L. TRIPP

The fashion industry traditionally has been centered in New York and Paris, but one Wisconsin inventor had a profound impact in the area of women's undergarments.

Freeman L. Tripp, born in Eau Claire in 1842, invented and patented several popular styles of corsets, hat pins, and other women's items. He also invented non-feminine items like a copper-toed boot and a wagon spring.

Unlike entrepreneurs like William Harley or Ole Evinrude, Tripp was interested in inventing and not in manufacturing the items he created. Although he made a considerable amount of money from his inventions, those who manufactured and marketed them made millions.

Tripp's interest in women's undergarments stemmed from the fact that he ran a women's wear shop in Madison in the 1860s. At the time, whale bone was used to stiffen hoop skirts, but Tripp invented a short-waist corset made of whale bone. Women of the era often wore corsets to reduce their waist size and in place of a bra, which hadn't yet been invented.

A Chicago firm was so impressed with Tripp's corsets that it pressured him to manufacture them, so Tripp sold the patent to a traveling salesman for $500. Later, Tripp also invented a long-waist corset and sold that patent for $500. He created a low-bust model and an abdominal corset.

Before Tripp invented the hat pin, women's hats were fastened with ribbon bonnet ties. His hat pin design became so popular that bonnet ribbon manufacturers felt threatened and tried to outlaw the hat pin. Tripp sold his hat pin stamping machine for $10,000, although manufacturers made millions of dollars from the hat pins they sold.

Before the Civil War, Tripp got the idea of nailing copper pennies to the toes of his boots to protect them, inventing the first metal-toed shoe. A friend put up the money for a factory and Tripp sold out his invention to him for $10,000.

Tripp also invented a spring wagon tongue that took the weight of a wagon load off the neck yoke so horses would have it easier in pulling the wagon. He sold this invention for $5,000 and the man who bought it made a fortune.

ROLF DARBO

By the time of his death in 1984 at age 77, Rolf Darbo had more than a dozen patents to his credit. He invented the Insty Grill, a portable outdoor grill for camping or picnicking; Temp Guard, a device placed in a window that signals a neighbor when a house becomes too hot or cold; and Nitro Tablet Dispenser, a device to dispense nitroglycerin tablets to people with angina.

He came up with many of his inventions in a recreation room at his apartment in the Chalet Gardens complex.

Darbo didn't invent a better bread box, but he did invent a similar item. He invented the Snack Box, a device designed to take advantage of cold Wisconsin winters. First, he charted Madison's temperatures, learning that temperatures are below 50 degrees for 5 1/2 months each year. He used a 25-watt bulb to keep food inside the box from freezing while the temperatures would keep the food cold. The idea was to avoid tromping through the snow to get a snack.

A real estate agent and owner of apartment buildings, Darbo found a ready market for his Snack Box among university students who were

Wisconsin State Journal archives

Madison inventor Rolf Darbo in his lab.

his tenants. He told the *Wisconsin State Journal*:

"The students like to have a cold can of beer or bottle of milk around while they're studying or some sandwich to munch on, but they don't want to go to the expense of renting rooms with cooking facilities. This isn't designed to replace a refrigerator. During the summer months, a fellow doesn't mind running over to the corner grocery for some snacks. It's those cold winter months that make him wish he had some kind of refrigeration."

While working on the Snack Box, Darbo got the idea for Temp Guard, which he saw as the ideal thing for snowbirds to make sure their furnace didn't fail and the pipes didn't freeze while they were basking in the Florida or Arizona sunshine. His first prototype of the device was simply a lamp placed in the window and attached to a thermostat. If the temperature fell too low, the lamp switched on. He tried to market a later version of the device to fishermen, saying they could encase it in a plastic bag and lower it into the water to check the water temperature. He called the later version Cold Cup.

Darbo's antifreeze tester was a more successful invention in the marketplace. He sold seven thousand of them before a corporation bought the rights from him for $16,000. The testers checked antifreeze in a car or truck to make sure it protected the radiator from freezing. Darbo believed the testers would be a valuable tool for gas station attendants, who pumped gas for motorists before self-service stations became the norm.

His success with the antifreeze tester

prompted Darbo to retire from the real estate business. But his retirement wasn't permanent. In his later years, he owned and operated the Forest Hill Crematorium in Madison, and he devised the Nitro Tablet Dispenser in 1982. A person suffering from angina could put a pen-sized gadget in his or her mouth, click it and a nitro pill would pop out under their tongue. Darbo told the *Wisconsin State Journal* his device was better than trying to open a pill bottle.

"You have to take the top off a little bottle, pull out the cotton, and fish out one of those little pills. That's fine if you have lots of time, but what do you do if you're driving in the dark and you have an attack?"

Before boxer George Foreman retired from the sport and began marketing outdoor grills, Darbo invented his Insty Grill. The collapsible grill burned newspapers, and he said that eight pages would cook a meal. Invented in the mid-1970s, an era of concern about energy conservation, the Insty Grill saved energy and could fit into a suitcase.

Other inventions included a round heart, which he said could be printed on stickers and placed on luggage, calling cards, and other personal items to announce that a man or woman isn't involved in a relationship and is available. He also invented the Microtome, which sliced specimens to be examined under a microscope.

He sold 30,000 gadgets that hooked into the

end of a vacuum cleaner and held a wet shoe that he said would dry in eight minutes. But 50,000 of them were unsold and thrown away. He also invented a collapsible clothes brush and a silicone material that he put between dishes in his camper so the dishes didn't rattle.

He once described his inventions as 1 percent inspiration, 9 percent perspiration and 90 percent marketing.

MARGARETHE MEYER SCHURZ

Margarethe Meyer Schurz didn't invent the concept of kindergarten, but she was responsible for importing the idea to the United States.

Born in Hamburg, Germany, in 1833, Margarethe Meyer and her sister, Bertha, met Friedrich Froebel when Margarethe was about 15 years old. The sisters were very impressed with Froebel's kindergarten program, and in 1851, Bertha opened her own kindergarten with her husband.

Margarethe married Carl Schurz and came to America, where she began

Wisconsin State Journal archives

Margarethe Schurz

using Froebel's ideas for her young daughter, Agathe. They settled in Watertown, and soon Margarethe had rounded up a group of neighborhood children. They played games, sang

songs and participated in other activities that prepared them for school. Like many early kindergartens, Margarethe's informal gathering was conducted entirely in German.

Word spread of the impressive achievements of the neighborhood children so Margarethe opened an official kindergarten, the first in the United States.

The Watertown kindergarten continued long after Margarethe's death in 1876 at age 43 in Washington, D.C. after years of health problems. Prejudice against Germans during World War I caused it to close.

By that time, however, kindergarten had become an established concept in the American educational system. When Elizabeth Peabody, a nationally known education expert, visited Magarethe's home in 1859, she was so impressed with Agathe's achievements that she had helped spread the concept nationwide.

The first kindergarten building was restored and moved to the grounds of the Octagon House in Watertown in 1956, a century after Margarethe established the first kindergarten.

A plaque commemorates Margarethe's importance in the history of the American educational system: "In memory of Mrs. Carl Schurz (Margarethe Meyer Schurz) Aug. 27, 1833 — March 15, 1876, who established on this site the first kindergarten in America, 1856."

CARL ELIASON

By the 1920s, the automobile had become an important part of American life. But those vehicles were only of limited use during the winter in Sayner, Wisconsin, where heavy snowfalls made it difficult to get around. So Carl Eliason of Sayner had the idea of building a motorized sleigh that could propel itself on the snow. In 1924, he built the world's first snowmobile.

It took Eliason two years to build the snowmobile in a small garage behind his house. He often went to Milwaukee by train to buy parts. When it was finished, the snowmobile was powered by a front-mounted outboard motor and had steering skis controlled by ropes. The engine was started first and gotten up to speed, then the skis were lowered to the ground and the snowmobile began to move.

Eliason patented his machine in 1927, calling it a Motor Toboggan. Over the next fifteen years, Eliason continued to refine his invention. He used both two-cylinder and four-cylinder motorcycle engines and eventually developed some snowmobiles that could seat four people. He sold about forty snowmobiles over the next fifteen years to hunters, utility workers, and outdoorsmen. Each machine, which sold for $350 to $550, was custom-built and few were built exactly alike.

Eliason preferred Excelsior and Indian motorcycle engines to the Harley-Davidson engines because the engine and transmission were in a single unit.

The snowmobiles were gaining worldwide notoriety and when buyers from Finland ordered two hundred machines, the Four Wheel Drive Company of Clintonville took over production and Eliason served as a consultant. The Finland deal fell through, but the U.S. Army bought one hundred fifty snowmobiles for use in Alaska. Several Russians visited the Clintonville plant

Carl Eliason & Co.

Carl Eliason and his first snowmobile.

to test the machines. They mounted a machine gun on the front and pretended to fire as they sped up and down the Pigeon River.

A brochure was printed to advertise the snowmobiles in 1940, and over the next seven years about three hundred machines were manufactured and sold.

After World War II, snowmobile sales declined and Four Wheel Drive's snowmobile division was acquired by a Canadian subsidiary. Production was moved to Kitchener, Ontario, where the first rear engine model was built in 1950. Although the smaller models were successful, Polaris snowmobiles took over the market during the late 1950s after Eliason's original patents began to expire. In 1963, the company was sold to the Carter Brothers of Ontario, and the last Eliason snowmobile rolled off the line a year later. By this time, Ski-Doos dominated the market.

HERMAN MUELLER

Herman Mueller, who grew up in Ohio but later settled in Milwaukee, began working with machinery at an early age. He's best known for inventing the magneto and he sold the patent on it for a mere $500.

Among his many other inventions, however, was the squirt gun, a weapon he called a "fluid gun" with hermetically sealed collapsible cartridges originally invented as a humane method for Milwaukee police to deter crime.

"Why I was always fixing people's clocks at the age of 12," he told the *Milwaukee Journal* in a 1935 article. "By the time I was 15, I had as many as forty watches to repair in the house at one time. I never learned the trade from anyone, but I was known as an expert watch repairer for year."

Mueller ran a jewelry store for a while until it was bought by the Evinrude company, and he worked for them for five years. He invented a clock without a pendulum that didn't tick and another clock with an alarm that sounded like a whisper instead of a loud ringing sound.

His first invention was an electric clock in 1895. He also invented internal combustion engines with fuel valves that didn't clog. From 1898 to 1910, he built and sold automobiles. He built a one-ton truck that was used for deliveries by a local grocer. Mueller was recruited by automobile entrepreneur R.E. Olds to move to Michigan and supervise research, but Mueller refused.

"I guess that was a big mistake, not connecting with a going concern instead of forming my own company," he later told the *Milwaukee Journal.*

Although Mueller claimed his automobiles were among the best at the time, they failed to survive in the marketplace.

But his biggest mistake may have been being outwitted on his invention of a magneto, a key electronic component of a gasoline engine, and selling the rights for a fraction of its value. Mueller said one of the pitfalls of being an inventor was keeping constant watch so someone didn't infringe on your patent. He eventually devised a method of photographing his inventions and mailing the photograph to himself. The postmark on the envelope would provide the date of the invention.

Despite his reputation as a genius mechanic, Mueller never made a fortune from his inventions although others profited mightily from them.

JOHN APPLEBY

When John Appleby carved out a farm from the wilderness near Beloit in the 1830s, he had to make some of his own tools. Like many pioneers, he had a particular knack for invention.

"Whatever was needful in the shape of a tool, or a machine, to accomplish desired ends, he possessed the peculiar inventive facility to produce it," wrote Waldo Stone, who knew Appleby when Stone was a boy and later became an investor in his invention, in a letter published by the *Beloit Daily News*.

Wheat was the principal Midwestern crop in the mid-1800s. Harvesting grain was a slow process for the pioneers, and one of the problems was that farmers used wire to bind the grain. Using wire did not allow them to use efficient machinery in the field, and the wire would damage mill stones or bits of wire would mix with the grain.

The Marsh harvester was an early attempt to mechanize the wire binders, and C.W. Marsh continued to produce wire binders even after it became clear that the twine binder had taken over the market. The *St. Paul Pioneer Press* editorialized in 1878 that wire binders were dangerous.

"The case is a serious one and every farmer who uses wire for binding should see to it that the bands are removed when threshing, else the price of the wheat may be materially reduced," the *Pioneer Press* said.

Appleby worked in a small shop on his farm trying to perfect a new way of binding grain in the field.

By the 1850s, Appleby invented the twine binder, a device that revolutionized farming by mechanizing the grain harvesting process. The invention was so successful that Appleby got the support of two investors, Parker and Stone, to begin manufacturing them. In 1878, 115 twine binders were sold throughout the Midwest from Texas to North Dakota.

Farmers raved about the new method of binding grain.

"When the grain stood up it made splendid work binding every bundle tight and leaving them square at the butts, and it handled the lodged grain better than any wire binder in the vicinity," wrote farmer Frank Nickerson in the *Beloit Daily News*. "In fact, wire binders are left behind in this vicinity, as everyone who saw this work preferred it to any other."

As the twine binder was perfected, it eventually could cut a swath fourteen feet wide and bind it into bundles that then would be taken to a mill to shake the grain loose from the bundles.

The twine binder along with the reaper and other farm machines pointed the way to modern agriculture in which farmers could work a vast number of acres without a lot of farm hands helping them.

"Every manufacturer of implements for the harvesting of grain was a part of that ponderous wheel of progress which has led the advance of civilization in this country and opened up the infinite resources of the land for the use of man," Stone wrote in 1911.

JEROME INCREASE CASE

Like John Appleby, Cyrus McCormick and Eli Whitney, Jerome Increase Case was an inventor and entrepreneur who transformed agriculture. He was the first to adapt the steam engine to farm use. He also would found a company that would evolve into a multinational giant.

As a young man, Case was intrigued by an article he saw in the *Genessee Farmer* about a wheat threshing machine. Before such machines were invented, wheat was harvested by hand with scythes and tossed into the air to separate the grain from the chaff.

Case lived in Williamstown, N.Y., until he was about 23, when he brought a crude threshing machine he invented with him to Rochester, Wisconsin. He started a company to manufacture the threshing machines and, a year later, moved the fledgling company to Racine and built a factory to take advantage of water power.

During the same era, Cyrus McCormick was establishing his reaper manufacturing business in Chicago. His invention of the reaper in 1831 also would help revolutionize agriculture, and McCormick's company later would become International Harvester.

The market for farm machinery boomed during the Civil War and the number of reapers and farm mowers rose from 90,000 to 250,000. In 1863, Case formed J.I. Case and Company with three partners: Massena Erskine, Robert Baker and Stephen Bull. An eagle trademark was adopted a year later based on the exploits of Old Abe, a bald eagle that was the mascot for the 8th Wisconsin Regiment and participated in many Civil War battles. It would serve as the company logo for more than a century.

Case produced his first steam engine for farm use in 1869. It was wheel-mounted, drawn by horses and used only to produce power. But it would be more than a decade before steam power became popular with farmers. In two decades, however, his company would become the largest producer of steam engines worldwide.

Racine Heritage Museum

Case's first steam engine.

When the great

Chicago fire destroyed McCormick's factory in 1871, Case offered to help McCormick build machines at his Racine plant. But McCormick refused and instead build a new factory on the southwest side of Chicago. McCormick also began production of the twine binder invented by John Appleby.

Racine Heritage Museum

Jerome Increase Case

A factor in Case's success was his devotion to customer service. On one occasion when a plant mechanic and a dealer could not repair a Case thresher on a Minnesota farm, Case visited the farm personally and tried to repair it himself. He got so upset that such an inferior product had come from his factory that he poured kerosene on the machine and set it on fire, then got a new thresher for the farmer.

Case also served three terms as mayor of Racine and two terms as a state senator. He was president of the Manufacturer's National Bank of Racine and founded the First National Bank of Burlington. He also founded the Wisconsin Academy of Science, Arts and Letters.

He died in 1891, and the company he founded went on to become one of the largest farm machine manufacturers in the world.

International Harvester was formed in 1902 by consolidating five competitors that included McCormick's company. The merger launched an era of fierce competition with Case and John Deere.

But Case had become the largest producer of farm steam engines and threshing machines, and in 1904 it introduced the first all-steel thresher. The company also had expanded to international markets, establishing a branch in Argentina.

The first gasoline tractor was sold in 1906 and Case soon established the Case Tractor Works near Racine. Both International Harvester and Case bought plow companies in 1919, adding products to their inventory. Within a few years they would be locked in tractor price wars with Ford also in the market. In 1924, Case produced its 100,000th thresher.

The company would thrive into the 21st Century, acquiring International Harvester's agricultural equipment operations in 1985.

SEYMOUR CRAY

If one person can be identified as the father of modern computer technology, that person may be Seymour Cray of Chippewa Falls. His development of super-computers from the 1950s to the 1970s paved the way for a personal computer on everyone's desktop. He's been described as the Thomas Edison of the super-computing industry.

He was born in 1925. His father worked as a civil engineer with Northern States Power Co. and later became town engineer in Chippewa Falls. Young Seymour was fascinated by electronics, and as a boy he built radios and electric motors. He devised a telegraph connection between his bedroom and his sister's bedroom so they could communicate in Morse code after their parents had sent them to bed. When his father complained about the clicking noises at night, the innovative boy converted the system to flashing lights.

When he graduated from high school in 1943, Cray served in the military during the final years of World War II. He was sent to Europe and fought in the Battle of the Bulge and later served in the Philippines.

Wisconsin State Journal archives
Seymour Cray

He earned a bachelor's degree in electrical engineering from the University of Minnesota in 1950 and a master's degree in mathematics a year later. He got a job with Engineering Research Associates, a technology company barely a year old. ERA had developed cryptographic equipment for the Navy. Cray set to work in a plant that had been used to build wooden gliders for the D-Day invasion.

Computers were little more than a concept when Cray began researching them. He worked on machines that later would evolve into the first Univac computers. When Remington Rand acquired ERA a few years later, the focus shifted to making the primitive computers more stylish and eventually the company decided to abandon the effort. Cray joined Bill Norris to help found Control Data Corporation.

At Control Data, Cray built scientific computers that were fast for their day. By age 34 in 1960, he had built several computers at Control Data. When he thought the company was becoming too large, he left the corporate environment and moved back to Chippewa Falls where he experimented with integrating transistors into computers. He built the first transistorized computer and invented a computer instruction language.

He continued to build computers for Control Data until 1972, when he started his own business, Cray Research Inc.

The first project of the new company was Cray 1, a computer shaped like a cone to shorten the wiring inside. He went on to build several more Cray computers that reached speeds of 500 megahertz. But Cray's focus on research and development instead of marketing eventually doomed the business to economic failure after he created Cray Computer Corporation. In 1994, the last computer he developed operated at a speed of 1 gigahertz, faster than any computer had ever achieved. Today's laptop and desktop computers are much faster than that.

"Seymour regarded every system he worked on as a stepping stone to the next," said Charles Breckenridge of SRC Computers in a tribute to Cray at a 1996 super-computing conference. "And most of them were foundations for other systems built by others using his basic designs. It is ironic that most of the competition for Seymour's machines came from companies that he had been instrumental in making successful."

Cray's biggest struggle was against what he called distractions. He had returned to Chippewa Falls to reduce distractions. To clear his mind, he enjoyed activities like skiing, tennis and wind surfing or digging his own tunnels. When someone realized he didn't have a phone in his office, they ordered one for him and asked where he wanted to put it. "On the tree outside my office," he said.

Cray died in 1996 of injuries he suffered in a car accident in Colorado.

BROOKS STEVENS

The world of Brooks Stevens might have been shattered when he was stricken with polio at age 8. Doctors said he probably would never walk again and he was unable to use his right arm.

But his father, William Clifford Stevens, wasn't willing to let the boy waste away in bed. He encouraged young Brooks to draw and build models of airplanes and boats. He taught the boy to ride a bicycle and promised to buy him a Model T Ford if he swam a mile in a pool.

"My father knew how to motivate me," he later said. "They must have hauled me out hundreds of times short of that mark. But I finally got that car."

That early motivation helped Stevens become one of the world's most prominent industrial designers. Among his many designs is the Oscar Mayer Wienermobile.

Stevens was born in Milwaukee in 1911. In 1929, he enrolled at Cornell University to study agriculture but left four years later without graduating. He worked as inventory manager for soap companies and a grocery supplier. But a glimpse of his future as a successful designer came when he won a logo design contest for his

Wisconsin State Journal archives

Brook Stevens leans on an Evinrude out-board motor he designed.

father's company and redesigned product labels for the grocery company.

Instead of moving to New York where the best known designers had offices, Stevens stayed in Milwaukee and opened an office in July 1935. In four years, the young company had 33 accounts and four staff members. He won new clients through lectures in which he talked about the importance of good design.

In 1937, he designed a house for himself and his new wife, Alice Kopmeier. Stevens and his wife raised four children and lived in the house for 50 years. During World War II, he seized upon the opportunity to convert military equipment to civilian use and converted the Army Jeep into a station wagon.

After the war, Stevens designed the Hiawatha train for the Milwaukee Road with a glass-covered observation car that allowed travelers to view the scenery they passed. By the early 1950s, he had amassed enough renowned designs for a display at the Milwaukee Art Institute. By 1952, products that he had designed were ringing up a billion dollars a year in sales.

His passion was cars and Stevens designed many of them, including the short-lived Paxton in 1953. He also designed cars for American Motors, Volkswagen, Alfa Romeo and Studebaker. He collected classic cars and displayed them at the Brooks Stevens Automobile Collection & Automobile Museum in Mequon. The museum closed in 1999.

Stevens wanted to design a car that was dependable and safe, but with a classic design. He came up with the Excalibur, which was similar to the Studebaker.

Besides cars, other designs included the Steam-O-Matic iron, tractors for Allis Chalmers, an automatic clothes dryer, the Miller High Life logo, motorcycles for Harley-Davidson, lawn mowers for Lawn Boy, boats for Evinrude, a tricycle and the Wienermobile.

When he went to Minneapolis for a speech in 1954, Stevens coined the phrase "planned obsolescence" to describe the idea of getting consumers to want more and better products. The concept, somewhat controversial at the time, became a key function of American manufacturing, influencing products from cars to computers.

By the time of his retirement, Stevens had designed more than three thousand products for six hundred clients. Besides the tenacity he developed by overcoming the ravages of polio as a boy, another key to Stevens' success was his willingness to adapt to the times. Asked at the end of his career whether he would have done anything differently, he replied, "Hell yes! Everything, because it's all outmoded."

He died in January 1995, at age 83.

Many of Stevens' designs were featured in an exhibition at the Milwaukee Art Museum during the summer of 2003.

DONALD DUNCAN

On a trip to San Francisco in 1928, businessman Donald Duncan watched Pedro Flores draw a crowd around him while performing tricks with his yo-yo. The crowd's enthusiasm persuaded Duncan that the yo-yo could be marketed nationwide. He bought out Flores and went to work producing yo-yos.

Despite the Great Depression economy that dashed the hopes of many entrepreneurs, Duncan's yo-yos served as an inexpensive diversion during the bleak era. He hired a cadre of yo-yo artists that he called the "Duncan Yo-Yo Professionals" to teach yo-yo basics and demonstrate tricks. Competitors soon tried to duplicate Duncan's success, but he trademarked the word "yo-yo" in 1932 so they were forced to use less popular terms like "come-back" or "whirl-a-gig."

The Duncan Company moved to Luck, Wisconsin, in 1946, making the small northern Wisconsin community "Yo-Yo Capital of the World." Duncan said he chose the location because it was close to the supply of hard maple he needed. The company produced about 3,600 yo-

yos an hour, using a million board feet of maple a year.

About forty-five million yo-yos were sold in 1962, five million more than the total number of the nation's children. But high advertising costs were taking their toll on the company's profits and in 1965, Duncan lost its "yo-yo" trademark, so competitors could use the name to describe their products.

In 1965, the Duncan company reached the bottom of its downward arc and declared bankruptcy. It was acquired by Flambeau Plastics Company, which continued to market about a dozen different models of the Duncan yo-yo. They were made of plastic and not hard maple.

Wooden yo-yos survived until 1972. They were made by Fred Strombeck, who bought Duncan's lathes in 1967 and marketed them under the Medalist brand.

On June 6, which is Donald Duncan's birthday, National Yo-Yo Day is celebrated in his honor.

ANSEL KELLOGG

Kelloggs are best known for cereal, but a different Kellogg made a crucial contribution to the newspaper business. Ansel Kellogg was the inventor of what was known as "patented insides." It was the idea that newspaper sections could be printed in advance. Today, newspapers around the world still make use of Kellogg's invention, stuffing pages containing the latest news with preprinted sections of advertising and editorial material.

Kellogg was the youngest of six children, born March 20, 1832, in Reading, Pennsylvania. His ancestry included John Rogers, who died as a martyr in 1665 under English Queen Mary. The family moved to New York City when Ansel was two years old, and his father worked as a wholesale and retail merchant.

After graduating from Columbia College in 1852, Ansel Kellogg worked as an apprentice to architect Richard Upjohn, but soon decided that his career should be in journalism. He came to Portage, Wisconsin, at age 22, where he landed a job with the printing firm of J.C. Chandler. Kellogg worked at a weekly paper called the *Northern Republic* and eventually became editor. He invented what was called a "job printing press" to help streamline production of the paper.

When the Civil War came, the Portage newspaper lost its compositor to the Union Army, making it impossible to produce the newspaper. But Kellogg came up with an idea then unheard of in the newspaper publishing business. He went to Madison and got the *Wisconsin State Journal* to print one side of the entire newspaper. The rest of the paper was printed back in Portage.

Kellogg's system of preprinting a portion of the newspaper became known under various names as patented insides, auxiliary sheets and ready print. He moved to Chicago in 1865, patented his invention and started a business to supply the nation's newspapers with preprinted material. He made a fortune before his printing equipment was destroyed in the Great Chicago Fire of 1871. When he got the presses rolling again, he quickly won back his newspaper customers who had gone to competing firms in the interim.

Kellogg had outstanding mathematical ability and a quiet disposition. Among his inventions was a system to add several columns of numbers at the same time. He also was a good writer who penned a book on his family tree.

During his later years, health problems persuaded him to make several trips to Europe seeking help. He died in 1887 at age 55, leaving his wife and two daughters. A large portion of Kellogg's fortune came to the Kellogg family in Baraboo, Wisconsin, where his brothers, Chauncey and Albert had settled decades earlier.

JOHN MUIR

John Muir's accomplishments as an environmentalist and author are well-known, but he also was an inventor beginning when he was growing up in Wisconsin.

Young John worked on his inventions when he awoke in the middle of the night, then worked all day in the fields of Hickory Hill Farm. He didn't have a lot of tools and once made his own saw from a strip of metal he took out of a corset. One of his first inventions was a clock that woke him in the morning.

Muir grew up in Dunbar, Scotland, until age 11, when he moved to the Wisconsin farm. He helped his family clear the forest, dig wells and plow fields with a team of oxen. He later wrote that the challenges of life on the farm prepared him well for his later wilderness adventures.

He invented locks for doors and windows on the farm. One time, farm workers waited in the rain for young John to return home with the keys so they could bring their threshing equipment back into the barn. He also invented a thermometer that was so accurate it fluctuated if a person was standing nearby.

At a schoolhouse, Muir devised a system to

light the fire before the children arrived. He later said: "I had only to place a teaspoonful of powdered chicrate of potash and sugar on the stove hearth near a few shavings and kindling and at the required time, make the clock by a simple arrangement touch the inflammable mixture with a drop of sulfuric acid."

When he went to study at the university, Muir's inventions continued. He invented a bed that propelled him out of it and onto his feet every morning.

"When the minutes allowed for dressing had elapsed, a click was heard and the first book to be studied was pushed up from the rack below the top of the desk, thrown open and allowed to remain there the number of minutes required," he later wrote. "Then the machinery closed the book and allowed it to drop back into its stall, then moved the rack forward and threw up the next in order."

Muir went to Canada during the Civil War, perhaps to avoid being drafted. While he was there, he lost an eye in a mechanical accident. He also lost the sight in the other eye for a time, but his sight in that eye eventually returned.

After returning to the United States, Muir's first wilderness trek was a thousand mile walk from Louisville, Ky. to Savannah, Ga. He went to Florida and San Francisco, then to Yosemite in California, where he worked as a shepherd and ran a sawmill. After he got married, he moved to Martinez, Calif., where he became a successful farmer growing pears and grapes.

During this period, Muir still traveled to Yosemite and Alaska and began lobbying for the preservation of Yosemite as a national park. His political activity led him to help found the Sierra Club as an organization to champion environmental causes.

Muir, who died in 1914, is best known as the father of the environmental movement. But his unusual inventions as a young man were early indications of his brilliance.

JOHN BACHELDER

John Bachelder, who was born in New Hampshire in 1817 but spent his later years in Milwaukee, probably was the single person most responsible for invention of a practical sewing machine. Unlike other Wisconsin inventors, Bachelder received little profit from his invention.

After growing up in New Hampshire, Bachelder moved to Boston where he became an accountant with a transportation company that did business on the Middlesex Canal. When a railroad to Manchester, New Hampshire, was built and killed the canal business, Bachelder started a business dealing in dry goods and importing silk.

One day he spotted a primitive sewing machine in a shop window and soon became obsessed with improving the device. In April 1848, he sold his interest in the importing firm and used the $12,600 he received to work on developing a better sewing machine. In earlier machines, cloth was vertical as it went past the needle, but Bachelder created a horizontal table for the cloth. He also invented a double lock stitch made with a horizontal and a vertical needle.

But Bachelder was the victim of others who sought to steal his invention and profit from it. William Baker, a clerk in a nearby button store, visited Bachelder's shop several times and was impressed by the sewing machine improvements. He got a friend who knew more about mechanics to visit the shop in disguise and study the mechanism. Baker and his friend, William Grover, patented their own machine that sewed with a loop stitch they'd copied from Bachelder's machine.

Other visitors included William Singer, who became the largest manufacturer of sewing machines in the world. The Singer company eventually merged with Grove & Baker, and Singer machines still dominate the market today.

In 1849, Bachelder hired a lawyer to file a patent on his improved sewing machine. But the lawyer did shoddy work and the patent failed to cover some of the machine's most important features. The lawyer had failed to include the double lock stitch in the patent application. That left the door open for Baker, Singer, and others to develop their own machines using some of Bachelder's techniques.

When Bachelder tried to sell his machine to Boston clothing firms, tailors feared the new device would threaten their jobs and several walked out, forcing the companies to abandon the Bachelder machines. The stitch produced by his machines was unique so it was impossible for the clothing firms to covertly use the machines.

Bachelder not only had exhausted all of his funds developing the sewing machine, but he also had borrowed several thousand dollars. Finally, he sold his invention for $4,000 to an agent who came in disguise from the Singer company. By the time Bachelder's patent expired in 1877, Singer was selling 260,000 sewing machines a year. On the day the patent expired, Singer expected competition and reduced the price of its machines from $60 to $30 each.

After failing in the sewing machine business, Bachelder tried selling cotton and woolen goods. He moved to California for a while, where he served as president of the Napa Public Library, before settling in Milwaukee.

Bachelder died without the fortune he might have earned from his sewing machine. Singer, on the other hand, left $15 million when he died.

GEORGE S. PARKER

George Parker, inventor of the Parker pen, spread his business worldwide by traveling to faraway places with a black bag of sample pens that he would give away to dignitaries and businessmen. On five trips overseas, he gave pens to hotel clerks in Pago Pago and distributed them in the Fiji Islands. He even gave an expensive jeweled pen to the king of Siam.

Parker was born on Nov. 1, 1863, in Shullsburg in southwestern Wisconsin. He later attended Upper Iowa and Valparaiso universities. The telegraph was one of the world's most important means of communication in the mid-19th Century. Parker became a telegrapher, and later teacher of telegraphy in Janesville, where he often was called upon to fix the pens his students used.

He became so proficient at repairing pens that Parker decided to try to make a better pen himself. In 1890, Parker and W.F. Palmer, a Janesville insurance salesman, patented their own fountain pen, and two years later they incorporated the Parker Pen Company in Janesville.

The fountain pen was unique because it carried its own reservoir of ink, making it more portable than pens that had to be dipped repeatedly into inkwells that sometimes were incorporated into the design of desktops. In 1894, Parker patented the "lucky curve" design for his pens. When the pens were upside down in a shirt pocket, the ink would drain back into the reservoir. When the pen was turned over and the cap removed, the ink would drain back toward the writing tip

The first product of the Parker Pen Company was the Lucky Curve Pen, which proved so popular that it would carry the company for three decades.

By the early 1900s, Parker had invented the Trench Pen, designed for soldiers to use in the field. They could mix water with black ink pills to produce their own ink. The U.S. War Department was the principal buyer of the Trench Pen.

By 1920, Parker had constructed a five-story manufacturing plant and headquarters for his company in Janesville, and annual sales had surpassed $1 million. During the 1920s, his travels spread the name of Parker pens across the world and he opened another plant in Toronto.

At the time, most pens were black but Parker began selling the Duofold, an orange fountain pen, in 1921. The pen was priced at $7, almost double the price of other fountain pens, but was successful. The Duofold was succeeded by the DQ (Duofold Quality) pen. Marketing was aimed at students, and the pen came in green, blue, and yellow. By this time, Parker was making his pens out of unbreakable plastic.

The company weathered the early years of the Great Depression, and in 1933 Parker introduced yet another pen design, the Vacumatic with had more ink capacity than the company's other pens.

Although Parker died at age 73 in 1937 in Chicago, his company continued to produce new pen designs into the 1960s including a self-filling fountain pen and liquid pencil. The company also marketed the Jotter, a quality ballpoint pen. In 1976, the company merged with Manpower Inc., a temporary worker agency based in Milwaukee, and was acquired by Gillette Company in 1993. In 1999, Gillette announced closure of the Janesville plant and the layoff of 300 employees.

Despite his company's worldwide roots, Janesville remained his home. He served as chairman of the Mercy Hospital board and donated $10,000 for a Janesville Women's Club building. He also supported Republican candidates.

CHESTER BEACH

When th U.S. Standard Electrical Works of Racine hired young farm boy Chester Beach in 1904, owner Fred Osius probably didn't realize that his company would be responsible for putting the gadgets into American life. Beach later invented the first high-speed and light weight motor that could run on either AC or DC power. This simple invention allowed the developed of hair dryers, vacuum cleaners, electric can openers, blenders, mixers, and other gadgets we take for granted today.

Beach, who became known as a natural mechanical genius, was born in 1880 on a farm in Ives Grove. When his wife, Ella Koenig, refused to live on the farm, the couple moved to Racine, where Beach used his inventive talents to land a job with Osius' company.

By 1905, Beach had invented a motor of less than one horsepower that could run at 7,500 rpm (revolutions per minute). In the process, he had to devise ball bearings that could withstand the heat and higher speed.

At that time, inventing a motor that could run on both alternating current (AC), the kind of power used in homes and stores, and DC (direct current), the kind of power derived from batteries, made the device compatible with the two competing kinds of electricity. Thomas Edison preferred direct current while some manufacturers used alternating current. Today, a device that runs on both systems, like a laptop computer or portable television set, can be used outside or in the house.

One of the Osius company's first products was a portable vacuum cleaner. In 1909, the McCrum-Howell Company of New York paid Osius $300,000 for rights to the vacuum cleaner, which was a small fortune at the time. Osius used the money to start another company, which he named after Beach, his mechanical genius, and Louis Hamilton, a cashier who had started at the company about the same time as Beach.

Hamilton was the business manager, Beach the product developer and Osius the financier. Osius paid his two employees a thousand dollars each to use their names, which he thought sounded better than his own. The new firm was Hamilton Beach, which became the leading manufacturer of household appliances.

Beach had invented his universal motor by 1910. It was used to produce mixers, fans, and knife sharpeners. The first product was a handheld massager followed by a drink mixer used to make milk shakes. At that time, malted

Wisconsin State Journal archives

Chester Beach

milk was prescribed by doctors as an aid to bodybuilding, and Beach was encouraged by the Horlick Malted Milk Co. of Racine to adapt his motor to make milk shakes. But the cake mixer may have been the company's most popular product. Sewing machines also became motorized instead of using a foot treadle to power them.

Hamilton and Beach left to form Wisconsin

Electric Co., which manufactured small motors used in power tools. The company's name was changed to Dunmore Manufacturing Co. in 1929 and the company still operates in Racine. But the Hamilton Beach company carried on without them and produced multiple spindle milk shake mixers used in soda fountains in the 1930s.

By the time Beach died in 1934, he had registered twenty-five patents. His universal motor powered a mechanical era in Racine in which the city became a mecca for small appliance development. By 1940, Racine had become known as the small motor capital of the world.

Hamilton Beach had produced more than thirty million electric motors by the early 1960s.

JOHN MELLISH

John Mellish grew up on a farm just south of Cottage Grove, Wisconsin, where his family grew corn, oats, and potatoes. During the growing season, he worked from dawn to dusk in the fields. On long winter nights, he watched the stars.

Mellish's fascination with the stars began when he got a small spyglass at about age 16. He turned the spyglass skyward and was surprised at the moon and stars he could observe. But the spyglass soon proved not enough for him, so he saved money he earned from helping his uncle with a carpentry job and bought a $16 telescope with a two-inch refractor. Soon that telescope, too, proved insufficient, but Mellish couldn't afford the $200 telescope he coveted. So, using two pieces of glass and emery powder, he began grinding his own lenses and made his own telescope.

"I sent to Chicago and secured a couple of pieces of plate glass," Mellish told the *Wisconsin State Journal.* "Out of one of these, I made a concave lens six inches across. I made it by grinding one piece of glass upon the other with emery dust in between."

In 1907, Mellish was 21 and he became famous when he spotted a faint object in the sky that looked to him like a tail of smoke. He had discovered a comet. The comet was named II Grigg-Mellish because another astronomer had seen it about the same time.

Mellish's homemade telescope was a curiosity in the area, and neighbors often dropped by to take a look at the stars through the contraption. Mellish wrote several magazine articles on how he built the telescope, which prompted letters asking him to build more telescopes.

He hoped to leave the farm and become a full-time astronomer. But that dream would take nearly a decade. In 1910, he was one of the first people in Wisconsin to see Halley's comet, and he wrote about it for the *Wisconsin State Journal.* When he discovered his third comet in 1915, Mellish got an invitation from Edwin Frost, the director of Yerkes Observatory near Lake Geneva, to spend some time at the observatory that summer. Mellish couldn't afford it, but managed to get a $300 grant to help pay for his stay.

A letter that Mellish wrote to Frost revealed his dedication to astronomy.

"I have been out all night several nights this winter," he wrote. "I am dressed to stand zero weather and do not feel the cold at all. I have sheep-lined shoes, so my feet did not even get cold and I am enjoying it."

Just before his arrival at Yerkes, Mellish an-

Cottage Grove Historical Society

John Mellish and his homemade telescope

swered a personal ad from a 17-year-old Illinois girl who said she was looking for the "perfect husband." Jesse Wood received two thousand letters in response to her ad, but selected Mellish. They were married and had eight children before divorcing in the 1930s.

Mellish spent well over a year as a volunteer research assistant at Yerkes, where he was assigned to search for comets. He worked with Edwin Hubble, then a graduate student, and E.E. Barnard, an older astronomer who grew up poor and was self-educated like Mellish.

His greatest discovery came during the fall and winter of 1915, when Mellish was studying Mars through a 40-inch refractor and suddenly was able to see the planet in much greater detail than ever before. He was the first astronomer ever to see craters on the planet, and he could see that some appeared several miles deep. Others were white-capped, perhaps with snow or limestone.

Mellish shared his discovery with Barnard, who pulled out drawings he had made in the early 1890s. He had seen dark areas on Mars that he thought might be forests. Although Mellish had made a monumental discovery, few

Cottage Grove Historical Society

An older John Mellish at Mount Palomar in Califorina.

people believed him because no one else saw the craters.

He soon moved on to an observatory at Leetonia, Ohio, where he also continued to earn money by making telescopes. Eventually, he moved on to Mount Palomar Observatory in California.

Mellish spent his life observing the stars and grinding lenses to make telescopes. His granddaughter, Lucy Knauss, who lives in Colorado, recalled visiting his shop when she was a child. He loved cats, and they would usually be full of dust from grinding the lenses. She said he would help her with math problems even though he only had a third-grade education and that he often talked about seeing the Martian craters. He often took family members on trips to places like the Grand Canyon and, she said, "He knew every bird and flower and rock and tree."

In the early 1960s, Mellish's papers and his drawings of the Martian craters were destroyed in a fire at his home in Oregon. Shortly afterward, Mariner IV flew past Mars in 1964, and photos from the satellite confirmed for the first time the craters that Mellish had seen nearly a half century earlier. No other astronomer had been able to see them.

John Mellish died at age 84 in a nursing home at Walla Walla, Washington. He never became as famous as other astronomers like Edwin Hubble, possibly due to his lack of education and the fact that he wasn't able to publish his findings. But his discoveries of comets and craters were significant accomplishments.

MORE INVENTORS & ENTREPRENEURS

Besides the inventors and entrepreneurs profiled here, Wisconsin spawned many others, including:

• **Edward P. Allis** acquired a small millstone shop called The Reliance Works and by 1869 the company was on the way to building its first steam engine. Allis renamed the company after himself and in 1901, a dozen years after his death, the company merged with three others to form to Allis-Chalmers Manufacturing Co. The company built its first farm tractor in 1914 and became one of the nation's largest producers of farm equipment before selling its farm equipment division in 1985.

• **Matthew Andis** of Racine, invented and marketed several models of electric hair clippers along with other inventions that included a device that warned drivers of low tire pressure and massaging equipment.

• **Peter Josef Frans Batenburg** of Racine, invented the four wheel drive truck and held more than eighty other patents.

• **Folkert Belzer** of Madison, invented a cold

storage solution to preserve human organs for transplant.

• **Linn B. Benton** invented a type unit that led to development of the Linotype machine used in printing until computerized typesetting replaced "hot type."

• **Valentine Blatz** opened one of Milwaukee's first breweries in 1851. A year later, he married the widow of his former boss. Blatz was the first Milwaukee brewery to expand nationwide, but the label was sold to Pabst in 1959.

• **Jim Boelke** of Neenah invented the Cat Dancer cat toy consisting of cardboard strips on a wire which eventually had annual sales of $2 million in 11 countries.

• **Eugene Butts** of Evansville, who started inventing at age 11, had at least eighteen inventions including a 1915 crate for ringing, castrating and vaccinating pigs; a washing machine; a portable saw mill; and a punching machine.

• **Catherine Taft Clark** and her husband, Russell, mortgaged their house in 1946 so they could rent a store with used bakery equipment. In 1972, they sold the Brownberry Ovens company for $5.5 million. Catherine Clark was born in Whitewater in 1907. Her father died when she was 8 and her mother supported the family by cleaning houses and running a laundry business out of her home. After graduating at the top of her high school class, she moved to Milwaukee to take a job at Schuster's Department Store.

• **James L. Clark** of Oshkosh invented a new process for making matchsticks during the 19th Century.

• **William Cameron Coup** of Delevan, invented the three-ring circus.

• The first Culver's frozen custard restaurant was opened in 1984 in Sauk City, Wisconsin, by **Craig and Lea Culver** along with Craig's parents, George and Ruth. **George and Ruth Culver** had operated several supper clubs and Craig Culver became fascinated with the fast food business after working summers at his parents' Farm Kitchen Resort in Baraboo, Wisconsin. Within two decades, the frozen custard chain of restaurants had expanded to more than 200 franchises throughout the upper Midwest.

• **Dexter Curtis** of Sun Prairie invented a zinc pad for horses in 1870 that was placed under their collars and prevented them from getting sore necks.

• **Augustine Davis** of Galesville invented an acetylene generator in 1904.

• **William Didier** of Racine, invented artificial limbs that bend at the knees and elbows after losing his leg in a work accident.

• **Albert Dremel** of Racine, built the first gasoline-powered lawn mower in 1921 while working at Jacobsen's Thor Machine Works. He founded his own company and invented electric erasers and a razor blade sharpener.

• **A.E. Freeman** of Eau Claire was said to be

an inventor of the incandescent light bulb, but was beaten in his effort to claim credit.

• **Nyal "Fuzz" Forstner** of Wisconsin Rapids, invented the metal freestanding fireplace.

• **Colonel H.A. Frambach** opened Wisconsin's first mill in Kaukauna in 1873 in which paper was made from wood pulp. The process was imported from Germany. Before that, mills in the state made paper from old rags and straw.

• **George Groton II** of Racine, designed a basket closure device in 1890 and a bridle bit.

• **Gunderson Tobacco** of Stoughton, started the coffee break so its female employees could go home and check on their families.

• **Earl Gunn** designed an automobile muffler for the Nash Motor Company of Kenosha that set a new standard for exhaust systems.

• **George Hallaver** of Two Rivers invented the ice cream sundae in 1881 at Berners Soda Fountain.

• **John Hammes**, a Racine architect, invented the first kitchen garbage disposal in 1927 and founded In-Sink-Erator to manufacture them in 1938.

• **Tracy Hatch** of Menomonie, invented the battery powered outboard motor in 1900.

• **William Horlick** of Racine invented malted milk in 1883 and founded the Horlick Malted Milk Co.

• **David Houston** of Cambria got a patent for

the first roll of camera film in 1881.

• **Manucher Javid** of Madison invented a urea solution in 1957 to relieve the swelling of tissues during brain surgery.

• **Warren Johnson** of Whitewater invented the electronic thermostat in 1883.

• **Jacob Jossi** of Dodge County invented pungent semisoft brick cheese in 1877.

• **Kimberly Klein**, 11, of Manitowoc, invented and marketed a holder for sunglasses that attaches to a car visor in 1997 after her dad kept losing his sunglasses.

• **Earl "Curly" Lambeau** was a shipping clerk earning $250 a month at the Indian Packing Co. in Green Bay when he and **George Calhoun** organized a football team in the newsroom at the *Green Bay Press-Gazette* building. Lambeau, who convinced his employer to put up money for jerseys, played on the team that became the Green Bay Packers and served as head coach for more than three decades.

• The **Lee-Metford** rifle used by the British army was invented in 1865 by a Stevens Point watchmaker named Lee.

• **Per Lynse** of Stoughton helped make rosemaling painting popular in the United States.

• **Max Mason** of Madison, a mathematician, invented submarine devices in 1927.

• **Oscar Mayer** started a Chicago meat market in 1883 with his brothers Max and Gottfried. The company, which eventually moved much of its operations to Madison, Wisconsin, was one of the first meat producers to receive federal approval in 1906. An advertising campaign launched in 1936 featured Little Oscar and the Wienermobile. The company was acquired by General Foods in 1981.

• **John Menard**, who grew up on a small farm near Eau Claire, Wisconsin, had trouble buying lumber on weekends for his construction business, so he decided to do something about it. In 1960, he started buying lumber in larger quantities and selling it to other area companies. The business expanded to more than a hundred home improvement stores in the upper Midwest.

• **Andrew Modine** of Racine invented radiators for farm tractors, oil coolers, and later was a pioneer in automobile air conditioning systems. He registered 120 patents during his lifetime, more than any other Racine inventor.

• **Charlie Nagreen** of Seymour, a meatball vendor, is among several people credited with inventing the hamburger in 1885.

• **John Oster** of Racine sold handheld hair clippers and after World War II, began marketing the blenders that made his name famous.

• **Frederick Pabst** was a steamship captain on the Great Lakes when in 1862 he met and married Maria Best, the daughter of Phillip Best, the owner of a Midwest brewery. Pabst soon became president of the brewery and increased production from five thousand to a hundred thousand barrels of beer a year.

• **Les Paul** of Waukesha invented the electric guitar and the eight-track recorder.

• **Edward Joel Pennington**, originally from Indiana, may have invented one of the world's first airships, which he supposedly flew from Racine to Kansas in 1897. But Pennington also was a well-known con artist who once built a car capable of running only long enough to impress potential investors. He left Racine in 1902 with an outstanding hotel bill.

• **Richard Raddatz** of Oshkosh tested an early submarine in the Fox River in 1897.

• **Edwin Reynolds** invented a 19th Century sewage centrifugal pump that powered municipal sewage systems in Chicago, New Orleans and other cities.

• **Joseph Schlitz** came to Milwaukee in 1850 from Germany and went to work as a bookkeeper for brewer and saloon owner **August Krug**. Schlitz eventually took control of the brewery and it would become one of the nation's largest.

• **Alton J. Shaw** of Milwaukee invented the electric crane used in construction projects.

• **Harry Soref** of Milwaukee invented the laminated padlock and worked with magician Harry Houdini.

• **Verner Soumi** invented satellite instruments to measure radiation from the earth's atmosphere.

- **Joseph Steinwand** of Colby invented mild Colby cheese in 1874.
- **John Stevens** of Neenah invented a metal roller to grind flour in 1872.
- **Daniel and George Van Brunt** of Horicon invented the broadcast seeder for planting crops in 1861.
- **S.R. Wagg** of Appleton invented a device to reduce the number of rags used in papermaking before wood pulp replaced rags as the raw material for paper.
- **Charles Warner** of Beloit invented the automobile speedometer and **Arthur Pratt Warner** was the founder of Warner Electric, which at one time provided the brake design for 75 percent of all mobile homes. Warner Trailer Co. built the first tourist trailer, ancestor of today's recreational vehicles.
- The **Wadewitz family** of Racine was involved in several manufacturing ventures ranging from making trunks to the Western Publishing Co., famous for the Little Golden Books series.
- **Towner K. Webster** of Racine started producing magnetos for small engines and later founded Webster Electric Co., producing electronic parts.
- **Samuel Whitesides** of Appleton invented a planing machine in the 1860s.
- **J.C. Wilson** of Appleton invented a rail car

coupler in the 1860s.

• **Elmer Winter** of Milwaukee founded Manpower Inc., a temporary job service company, in 1952 with **Aaron Scheinfeld**, his brother-in-law. Winter's initial investment was $7,000 and the company grew into a business with annual revenue of $12 billion and nearly three million workers in more than sixty countries. Winter, who turned 91 in 2003, has remained active in local politics.

• **Steve Wittman** of Fond du Lac invented spring landing gear used by Cessna.

• **Ron Wozella** of Plymouth invented action decoys for hunters.

• **Frank Lloyd Wright** of Spring Green invented the wall-mounted toilet.

• **John L. Wright**, son of Frank Lloyd Wright, invented Lincoln Logs.

• **Otto Zachow** and **William Besserdich** of Clintonville invented the first four-wheel-drive automobile.

BIBLIOGRAPHY

"60 Years of Industrial Design," Brook Stevens Automobile Collection Inc., *www.excaliburclassics.com*.

Bellis, Mary, "The History of the Yo-Yo (or what goes up must come down)," About Homework Help, *http://inventors.about.com*.

Bjork, Kenneth, "Ole Evinrude and the Outboard Motor," Norwegian-American Historical Association, NAHA Web.

Breckenridge, Charles W., "A Tribute to Seymour Cray," Charles W. Breckenridge, 1996, *http://www.cgl.ucsf.edu*.

Buenker, John D., Ammann, Richard E. and Sasso, Robert F., *Invention City: The Sesquicentennial History of Racine*, Racine Heritage Museum, July 1998.

CareerJournal.com, The premier executive career site from the *Wall Street Journal*.

"Case IH History," Case International Harvester, *www.caseih.com*.

"Eliason Snowmobiles," Carl Eliason & Co., *www.eliason-snowmobile.com*.

Fatica, Vincent, "George S. Parker," Syracuse University.

"Harry Miller: Automotive Genius," Dunn County Historical Society, *discover-net.net*.

"Houdini Biography," *MagicTricks.com*.

"Industrial Strength Design: How Brook Stevens Shaped Your World," Milwaukee Art Museum, 2003, *www.mam.org*.

"John Wesley Carhart," *The Handbook of Texas Online*, The Texas State Historical Association.

Jones, Del, "Johnson family legacy finds layers of love, loss," *USA Today*, Dec. 4, 2002.

Lapham, Increase Allen, *The Antiquities of Wisconsin,* University of Wisconsin Press, October 2000.

"Margarethe Meyer Schurz 1833-1876," Froebel Web, *froebel.50megs.com.*

"Ole Evinrude: Outboard boat motor," *Invention of the Week,* The Lemelson-MIT Program's Invention Dimension, January 1999, *http://web.mit.edu.*

Pepper, Jason, "Seymour Cray," *http://ei.cs.vt.edu/ ~history/Cray.Pepper.html.*

Rafferty, Tod, *Harley Davidson: The Ultimate Machine,* Courage Books, March 2002.

Rowland, Pleasant, "How We Got Started," *Fortune Small Business Magazine, www.fortune.com.*

Sahlman, Rachel, "Harry Houdini," Spectrum Biographies, *www.incwell.com.*

Schuldt, Gretchen, "Movie did not shine, Johnson Wax alleges," *Milwaukee Journal-Sentinel,* Dec. 8, 1999.

"Seymour Cray," Cray Inc., *www.cray.com.*

Wisconsin Historical Society Digital Library and Archives, articles published by various Wisconsin newspapers including the *Appleton Times-Crescent, Beloit Daily News, Milwaukee Journal, Milwaukee Sentinel, Oshkosh Northwestern* and *Wisconsin State Journal.*

INDEX